Key Concepts for Critical Infrastructure Research

Jens Ivo Engels
Editor

Key Concepts for Critical Infrastructure Research

Editor
Jens Ivo Engels
Darmstadt, Germany

ISBN 978-3-658-22919-1   ISBN 978-3-658-22920-7 (eBook)
https://doi.org/10.1007/978-3-658-22920-7

Library of Congress Control Number: 2018950043

Springer VS
© Springer Fachmedien Wiesbaden GmbH, part of Springer Nature 2018
This work is subject to copyright. All rights are reserved by the Publisher, whether the whole or part of the material is concerned, specifically the rights of translation, reprinting, reuse of illustrations, recitation, broadcasting, reproduction on microfilms or in any other physical way, and transmission or information storage and retrieval, electronic adaptation, computer software, or by similar or dissimilar methodology now known or hereafter developed.
The use of general descriptive names, registered names, trademarks, service marks, etc. in this publication does not imply, even in the absence of a specific statement, that such names are exempt from the relevant protective laws and regulations and therefore free for general use.
The publisher, the authors and the editors are safe to assume that the advice and information in this book are believed to be true and accurate at the date of publication. Neither the publisher nor the authors or the editors give a warranty, express or implied, with respect to the material contained herein or for any errors or omissions that may have been made. The publisher remains neutral with regard to jurisdictional claims in published maps and institutional affiliations.

Printed on acid-free paper

This Springer VS imprint is published by the registered company Springer Fachmedien Wiesbaden GmbH part of Springer Nature
The registered company address is: Abraham-Lincoln-Str. 46, 65189 Wiesbaden, Germany

Contents

Introduction 1
Jens Ivo Engels

Criticality 11
Kristof Lukitsch, Marcel Müller, Chris Stahlhut

Vulnerability 21
Stephanie Eifert, Alice Knauf, Nadja Thiessen

Resilience 31
Ivonne Elsner, Andreas Huck, Manas Marathe

Preparedness & Prevention 39
Arturo Crespo, Marcus Dombois, Jan Henning

Relations between the Concepts 45
All Authors

Bibliography 53

Introduction

Jens Ivo Engels

This publication is one of the first results of joint work in the interdisciplinary graduate school 'KRITIS. Kritische Infrastrukturen: Konstruktion, Funktionskrisen und Schutz in Städten' – 'Critical Infrastructures: construction, functional crises and protection in cities'. The graduate school is based at Technische Universität Darmstadt and started in autumn 2016. This book has been written jointly by all of the 12 doctoral researchers of the 'first generation' in that school and one of the supervisors. It is the fruit of our intensive discussions in the joint weekly *Oberseminar* and of several writing sessions. We think it represents a good argument for interdisciplinary collaboration as it proves that discussions and debates between humanities, social sciences and engineering researchers are possible – and fruitful. Every single author of this text has been confronted with cultures of thought that are foreign to her or his own discipline. That did not change our respective way of doing research. In many cases it was just the opposite in fact: becoming familiar with other research strategies helps clarify and consolidate our own discipline's approach. However, every author has learned to see the different disciplinary usages of the concepts elucidated in this text, and to ponder their strengths and weaknesses. He or she now will use one or several of the concepts in their own doctoral research with these insights in mind, and will be able to delimit its use from possible alternatives. In some instances, however, he or she will borrow insights from neighbouring disciplines.

The book has been published technically as an edited volume, for pragmatic reasons. In fact, literally every author has contributed to much more than his or her individual chapter. We consider the book as our joint work – not only the last chapter.

In alphabetical order, the book is authored by the following doctoral researchers: Arturo Crespo (Civil Engineering), Marcus Dombois (Civil Engineering), Stephanie Eifert (Medieval History), Ivonne Elsner (Urban and Infrastructure Planning), Jan Henning (History of Technology), Andreas Huck (Urban and Regional Geography), Alice Knauf (Political Science), Kristof Lukitsch (Contemporary History), Manas Marathe (Architecture and Urban Planning), Marcel Müller (Philosophy), Nadja Thiessen (Contemporary History) and Chris Stahlhut (Computer Science).

The chapters have been prepared, after joint discussions, by writing groups, observing the rule of bringing together different disciplinary perspectives. Criticality has been written by Lukitsch, Müller and Stahlhut, Vulnerability is authored by Eifert, Knauf and Thiessen, Resilience has been the task of Elsner, Huck and Marathe, and Preparedness & Prevention has been taken up by Crespo, Dombois and Henning. All of them contributed parts to chapter 7. Finally, Jens Ivo Engels (Modern and Contemporary History) took up the chapters, revised, complemented and merged them into one text and wrote the introduction.

The authors have benefited from the opportunity to discuss their ideas and their texts with several persons. We are thankful to Sabine Höhler, who commented intensively on a draft version when she was at Darmstadt as a KRITIS Mercator Fellow, Markus Lederer, who stimulated the discussions of the criticality redaction group, and Leon Hempel, who contributed to innumerable oral debates and textwork sessions when residing as a KIVA guest professor in Darmstadt. We continually profited from ideas and suggestions by the guests of the weekly KRITIS *Oberseminar* in 2017, especially Jörn Birkmann, Martin Endreß, Stefan Kaufmann, Richard Little, Benjamin Rampp, Christoph Riegel, and Erik van der Vleuten. Last but not least we thank the PIs and PhD supervisors of KRITIS graduate school.

Above all, we would like to highlight the decisive contributions of Jochen Monstadt to the configuration of the four concepts as a systematic approach to critical infrastructure research, as well as his inputs during several discussions over individual concepts. Finally, Tina Enders took care of the editing and publishing process.

1   Background and Aim of the Text

The work of the abovementioned KRITIS graduate school is organised into three research areas, namely the emergence or construction of critical infrastructures, function failures or problems, and strategies to protect these infrastructures against failures or to cope with failures. In the methodological part of our research programme, which is authored by Jochen Monstadt and Jens Ivo Engels, we have identified four key concepts (we call in German 'Brückenkonzepte'). The key concepts should provide to all scholars involved a kind of common language. These concepts are: first, criticality, being the defining feature of those infrastructures which are analysed by the group, second, vulnerability as a tool that may help assess reasons for the failure of infrastructure, third, resilience as a concept which may help explain why certain infrastructures do not fail, and, fourth, preparedness & prevention, initially conceived as two different strategies to act against failures. On the following pages, however, the reader will notice that we have revised some of these initial assignments and delimitations between the concepts. Indeed, the research programme did not provide specific determinations or definitions of the concepts. This text could serve as a kind of specification.

We are fully aware of the fact that our research and the abovementioned concepts are part of a big conceptual field. There are other concepts and debates which are closely related to our four notions – such as 'coping', often used with respect to the capacities of the so-called social infrastructures, 'precaution', which might be used instead of preparedness & prevention, 'robustness', and many others. However, we are convinced that our set of four concepts provides a framework to help analyse the nature of critical technical infrastructure and strategies to cope with failures in the sense of a necessary minimum requirement.

Our goal is to initiate cross-disciplinary consultation processes. We do not in the first place adhere to the somewhat utopian idea that interdisciplinarity could be merged into something like a transdisciplinary meta-science. It is not only our methods, but of course also our research questions and problems that are so different, and sometimes incommensurable. We do not intend to even out these differences, because we contend that every single set of problems and methods is legitimate and important. This is the reason why we do not single out one specific usage of each of the concepts presented here. Rather, it is all about knowing and naming differences.

That is how we understand our text. Of course, it does not present new scientific results, as our research is just in its initial phase. Thus, the reader will find a kind of condensed introduction to the meaning, the research and problems of each concept. Discussing the concepts, we aim to outline very briefly the main debates about these concepts and provide a common basis for debating them further. This will be far from comprehensive, as many of these debates are so important that the books and papers on them fill whole libraries. We are brief in assessing what we consider the most interesting parts in the state of the art, without neglecting lacunae and open questions. No lacuna will be filled, no open question be answered, however. We try to organise a written dialogue between the concepts and the disciplines, rather than fixing concepts 'once and for all'.

However, our work is informed by our common research interest in a specific set of empirical phenomena, namely urban networked technical infrastructure. Every concept will be treated mainly from the perspective of infrastructure research, although they are all applicable to a wide range of other phenomena, social and technical. As a result, readers will find in every chapter a statement on what we consider important and necessary concerning the shape and the content of the concepts. Therefore, this book goes beyond a neutral reproduction of the ongoing debates. We intend to delimit the handful of meanings and uses we consider fruitful for our research. Alongside the empirical focus on infrastructure there is a second factor influencing our choices, namely our science and technology studies perspective.

One of the challenges we are confronted with is the highly normative and popular character of many, if not all of our concepts. 'Resilience' for instance has become, in recent years, a term used in an inflationary manner not only by planners or politicians, but also by journalists, therapists and even in private talks to denote more or less any kind of strength or perseverance of a person, a society or a technology under stress. Their popularity makes these concepts socially and politically relevant. However, the effect is that the content of these concepts becomes vague, as many different meanings are connected to the terms. That can be compared to the career of the concept of 'sustainability', which is also a concept that originated in the emerging field of forestry science in the 18th century, was taken up by ecology and economy in the 20th century and has been transformed into a policy tool since the 1990s. The lack of preciseness, however, offers the advantage of bringing together different epistemological or basic assumptions. The popularity of many of these concepts also offers the opportunity to critically assess the political or normative backgrounds of their users.

From our academic standpoint, of course we will try to highlight shortcomings due to normative reductions in these debates. For instance, we will plead not to limit our understanding of 'criticality' or 'vulnerability' to purely negative contents; they are not confined to denote deficiencies but also offer capacities or productive dimensions. For instance, high vulnerability is associated with a bad state of affairs, and lower vulnerability with better conditions. However, Wiebe E. Bijker et al. (2014) try to explore the potential of vulnerability as a positive opportunity.

The same is true for positive concepts such as preparedness: preparedness measures may inevitably lead to new vulnerabilities (in reducing others). These choices should be made explicit to the reader – as well as any choice made when using one or several facets of the vulnerability concept. However, our aim will be to elucidate these concepts as analytical tools.

2 Basic Choices and Concepts

This text is about key concepts in infrastructure research. There are however some notions and some empirical and theoretical choices we made that we should explain beforehand.

2.1 Networked Technical Infrastructure

Our joint research is devoted to networked technical infrastructure. That means we exclude a huge range of infrastructures from our research, namely all what could be called social infrastructure, like for instance education, health care, banking systems, public media, public administration and so on. Of course, they all can be defined as critical infrastructures with good reasons – in most policy concepts regarding critical infrastructure protection they are listed, because they provide vital services in modern societies. We focus however on networked technical systems. This is not to exclude social, political or governance aspects, but to make sure that we concentrate on technology–society interaction. There can be perfect studies on the health-care system, for instance, which only consider administration and economy, without focusing on energy supply for hospitals. We make clear that we are interested in the technology–society intersection by focusing on the most important systems like transportation, energy supply, waste management and information technologies, which all rely on material grids.

Moreover, it is the network aspect we are interested in. Therefore, we focus on grid-based phenomena. (Technical) networks have specific features, like the interplay of nodes and edges, causing special challenges, like the so-called cascading effects. There is no doubt that they inherently have a systems character. Technical networks are also determined by social factors such as power relations, regulations and values and vice versa. We therefore take on a socio-technical systems perspective on these networked infrastructures (cf. infra).

2.2 Science and Technology Studies Perspective

In the last section, we insisted on the interrelationship of technical and social aspects. In fact we contend that there is no line between the two. Attention to these relations has been called by a scholarly movement named 'science and technology studies', or 'science, technology and society' (STS) that emerged in the course of the 1970s or 1980s. In short, this branch of research aims at analysing how social, political and cultural values, norms or practices affect the material world of technology and, in turn, the other way round. One of the key ideas in this research was the idea of the 'social construction of technology', underlining that technology does not develop in a separated world far from human society, but reflecting social, political or economic needs, interests or power relations (Bijker et al. 1987). In the field of empirical research, the conception and analysis of 'large technological systems' or 'socio-technical systems' was an important milestone, showing the complexity of how economic, material or physical, social, political and other factors interact when a technological solution prevails (and evolves) over time. According to Günter Ropohl (2009: 141), a socio-technical system integrates human and non-human technological subsystems into one unity of action. Another term used in this sense is that of 'networked societies' (Hughes 1987, Hughes/Mayntz 1988, Vleuten 2006).

The concept of STS has influenced how we consider the evolution of technology. Dissociating more and more from one-sided (social) constructivist approaches, research has privileged more and more the notion of 'co-construction' of technical systems and society, in order to mark a more balanced approach: society not only influences technology, but technology and the material world heavily influence society (cf. for instance Dolata/Werle 2007; Schueler 2008). Some prefer the notion of co-evolution or even co-production, highlighting the non-intentional but still mutual causal character of these relationships and processes (Murmann 2003; Jasanoff 2004).

In a similar way, actor-network theory and symmetrical anthropology, both heavily influenced by the work of Bruno Latour, have paved the way to consider non-human factors, including technology and technical devices, as forces of their own right, shaping for instance everyday human practices (Latour 1993). Based on this tradition, the properties of the physical world and 'material' factors in social life have been more and more integrated in their research by social sciences and humanities scholars (cf. Schatzki 2003). It is this symmetrical position we consider most fruitful to assess the reality of critical infrastructures in cities, conceived as socio-technical systems (cf. for infrastructures in cities Monstadt 2009).

However, the STS approach does not mean that we are restricted to considering 'the whole world' including the entire society and all technological arrangements. Rather, we know their interdependencies but focus on the grid, on the grid and its administration or operating institutions, on the relationships between the city as a social unit and its infrastructures, etc. In other words: we prefer a systems perspective.

2.3 Key Concepts are 'Boundary Concepts'

This text is about concepts, in other words about frameworks, which allow us to name and to conceptualise what we are researching, parting from different academic cultures and epistemologies. Our four key concepts are strategic 'bridgeheads' from which we intend to build bridges between the disciplines, because we are convinced these concepts are useful and understandable in each of the disciplines concerned. From a theoretical point of view, we consider criticality, vulnerability, resilience, and preparedness & prevention as 'boundary concepts'. This notion is inspired by a paper written by Stefan Kaufmann, who qualified resilience as a 'boundary object'. Kaufmann's position goes back to a famous article by Susan Star and James Griesemer, who depicted material objects (natural museum exhibits) as focal points for interdisciplinary understanding, allowing the transgression of epistemological boundaries (Kaufmann 2012; Star/Griesemer 1989). As our concepts are not material objects, however, we prefer the notion of 'boundary concept' (Mollinga 2008).

What is the nature of a boundary object or concept? It is a concept which is known and used by different institutions, stakeholders and above all by different scientific communities, because they find some common ground in it. At the same time, the concept is vague enough that each stakeholder, institution or scientific community may use it in a slightly different way, so that it is in conformity with the usage, interests, methods or epistemology of that community. Its vagueness is not a problem, but the very foundation of the common ground. Little misunderstandings between the different users of the concept are unavoidable and necessary. Each participant of the discussion thinks that everyone agrees with his or her individual understanding of the term, but actually each has a slightly different understanding. Following Kaufmann, these misunderstandings are extremely fruitful, because they enable interdisciplinary dialogue. Otherwise there would have been no chance to come together because of differing epistemological approaches.

This position is inspired by the conviction that differing scientific cultures are almost incommensurable. In a certain sense, scholars must trick themselves in order to build a common ground for communication.

If this is the case, every scholar should be aware of it. Of course, this is no legitimation to be unclear about the concepts. We are convinced that every researcher should be as explicit as possible and make his or her usage of the term as transparent as possible to the reader. This is a pledge not to unify the way in which the term is used, but rather to resist the use of ambiguous conceptual structures, while knowing that a rest of residuum misunderstanding will necessarily remain and will be fruitful.

Precisely, boundary concepts allow a cross-disciplinary, holistic view of infrastructures as socio-technical systems and of cities as socio-ecological systems.

2.4 System

During discussions and the preparation of this book we have noticed that the notion of 'system' is crucial; it almost emerges in the context of every key concept we are dealing with. Or, to put it the other way round, every single concept is conceived as applying to one or several systems; we consider them as systems-related. There are several reasons for our concerns with systems.

'System' is a concept which allows us to analyse both technical artefacts and human society, conceived either separately or as mixed phenomena. An electricity grid can be seen as a purely technical system, focusing for example on its physical and technical components and their relations. The political system of a country or city can be described as a purely social (or political) system concentrating on political actors and their relations. Finally, a city can be seen as a mixed system where socio-political elements are interlinked with technical components. So the very notion of system enables us to think of technical networked infrastructure as a matter of interaction between humans *and* artefacts. We therefore often use the term 'socio-technical systems' (cf. supra).

Second, our focus on infrastructure and cities makes it necessary to consider them occasionally as discreet entities. Of course the notion of system is not a container. Rather, it helps us model these entities in a specific way. With respect to technical networks and cities, that implies above all the idea that these 'systems' have certain 'inner workings': technical infrastructure, for example, is characterised by flows of goods, materials or energy. The political system of a city consists of many different actors ranging from users and voters to experts, administrators and the mayor. These components interact with each other, again according to certain logics.

As for networked infrastructure, much emphasis has been put on features like the so-called cascading failures or normal accidents, two notions that try to conceptualise certain dynamics in the case of function crises where complex yet interdependent technological bodies interact (Bak/Paczuski 1995). Concerning cities, there are also many concepts trying to characterise their function and the social and political processes within them. One of these concepts is called 'Eigenlogik', put forward by Helmuth Berking and Martina Löw (2008). This concept is similar, but not synonymous, to 'path dependency' – both concepts try to describe the limitations of possible developments as resulting from the tradition and the specific momentum in a given context, be it a city or be it a technical system (Beyer 2006: 14-21, 260).

Third, we can differentiate between the components on the one hand and the whole system on the other. In several chapters we will come back to this, because, for instance, criticality can be applied to a single component (or a group of components) as well as to a whole system or subsystem. The same holds true for resilience or vulnerability. We can focus on specific elements of a system, on relations between specific elements, or on the properties and implications of the system as a whole. In any case, the systems approach assumes that there are interrelations and interconnections between components and subsystems, regardless of the focus.

However, we do not adhere to any single systems theory nor do we propose a formal definition. We use the term in a pragmatic manner according to its abovementioned qualities. Again, the concept of system may be considered as a boundary concept including and enabling minor misunderstandings, fostering interdisciplinary communication.

2.5 Perturbation and Breakdown

We are interested in technical infrastructure; their very function is of course to operate and to provide certain services to society and/or to other technical systems. This is expressed by the notion of 'critical infrastructure' or by their 'criticality', denoting society's expectations towards technical systems (cf. infra). No city, no state, no private company would pay for a technical system that does not fulfil certain functions. Apart from this pragmatic justification, there is of course a theoretical one. The abovementioned systems perspective not only enables us to take into account both technical and social factors and their interplay, but also implies functional interaction. At the same time, we know perfectly that, empirically, most (technical) systems are very much characterised by the presence and ubiquity of problems, accidents and function breakdowns (Perrow 1984). Nonetheless, our idea of technical infrastructure is based, in the first place, on the ideal-type assumption that it does function and that it does provide the services it is meant to deliver. We take very seriously the social intentions inscribed into networked infrastructure.

However, this first-place assumption is the basis for analysing how and why operation breakdowns occur. Without the conception of 'normal operation' there can be no conception of operation breakdown. In our research, much is centred around the question of how societies or engineers react in the face of actual or potential breakdowns or operation interruptions. Every single concept we will elucidate in this book only makes sense in this context. However, we do not intend to theorise further the notions of breakdown or interruption. They will be recurrently used without further definition. It should only be clear that 'normal operation' is of course dependent upon social expectations – and the same is true for breakdown. Different social groups or stakeholders may conceive them in different ways. Whereas politicians and the media decreed a catastrophic breakdown in the course of the trans-European electricity blackout of 2006, engineers praised the successful containment of a potentially dangerous situation, proving the stability, not the instability, of the transnational power grid (Vleuten 2017).

3 Structure of the Text

The text is organised around the four key concepts. There is one chapter devoted to each concept, followed by a section detailing the relations between the concepts. A general bibliography concludes the book.

The chapters are organised in a similar structure. They are divided into four sub-chapters, namely (1) Ad-hoc Description, (2) Literature Overview, (3) The Concept within Critical Infrastructure Research, (4) Open Questions and Problems.

However, we start with a short introduction which provides a working definition or, if definition is not applicable, a starting point of understanding for the specific concept. As a rule, we tried to identify something that could be called the 'lowest common denominator' in the first place.

Subsequently, we try to describe the state of the art, cite some important contributions or detail how a term is used in different academic fields and traditions. The literature overview is also a sub-chapter where we typically try to make transparent differing forms of use of the concepts. Throughout our text we are very much concerned with structuring and explaining diverging usages of the concepts.

Sub-chapter 3 is systematically devoted to infrastructure research. Here we will make clear in which sense we think a certain concept is helpful to our ends. If ever we single out one or more perspectives that we consider the most fruitful, that will happen in this sub-chapter. As explained above, however, we will not commit ourselves to one single use of a concept. Again, this sub-chapter will give us the opportunity to describe different uses or perspectives a concept does provide. In the next section, we indicate some of the most important questions left open or to be discussed in the future. As we are dealing with a set of four concepts, which are interlinked in many respects, we systematically reflect upon these linkages in the last chapter. It is divided into six sub-chapters so that each concept can be paralleled with each other.

There is no general conclusion at the end of the book: we do not present definite results. Instead, we propose proceeding with the use of the concepts in empirical research.

Criticality

Kristof Lukitsch, Marcel Müller, Chris Stahlhut

Everyone who does research on technical infrastructures has to deal with a huge debate on the significance and weight of the so-called 'critical infrastructures'. Some would argue that this expression is a tautology, because infrastructure is by its very definition crucial or vital for a society. Technical infrastructure is conceived as forming the basis of the modern technological world (Laak 2004: 56). This applies in particular to the modern city – there is no modern city without technical 'infra-structure' at its bottom.

However, the concept of criticality is much more diverse and multi-faceted than that and it is worth considering it in more detail. Compared to the other concepts in infrastructure research elucidated in this publication, criticality has been more or less neglected so far in the academic debate. The concept of criticality is used in slightly different ways within different branches of research. Therefore, in this chapter we try to grasp criticality as a concept of reflection, trying to highlight the common elements of its use.

1   Ad-Hoc Description

Critical infrastructures are usually identified in one of two ways. The first one consists of simply enumerating all vital infrastructures, e.g. the electrical grid, transportation, administration, etc. This is the case in the US strategy for the protection of critical infrastructure of 2003, one of the first documents of this kind (United States Government 2003). Although enumeration is a common approach, we do not think it appropriate for scientific research.

We prefer the second approach, which defines criticality as a result of specific features. It ascribes certain (relational) properties to a system: a given system is critical in relation to other systems or entities. A system is critical for another one, when the former is necessary to keep running the latter. The German national strategy for the protection of critical infrastructure puts it like this: criticality is a relative measure for the consequences of a perturbation or a function failure concerning the supply of goods and services to society (Bundesministerium des Innern 2009: 7).

© Springer Fachmedien Wiesbaden GmbH, part of Springer Nature 2018
J. I. Engels (Ed.), *Key Concepts for Critical Infrastructure Research*,
https://doi.org/10.1007/978-3-658-22920-7_2

Critical infrastructure is thus infrastructure which is needed to keep running other major technical and/or social systems or which is needed to provide goods or services that are considered vital to the functioning of modern society. In this sense, Matthew Egan proposes a concise definition. Systems are critical if they '(a) provide routine functions [...] (b) no handy, rapid substitutes exist [...] (c) sudden dysfunction [...] causes nontrivial harm, and (d) they are embedded in [...] integrated systems' (Egan 2007: 5).

These two definitions show one of the major cleavages in criticality research, namely the difference between systems-based and consequence-based application (cf. below).

2 Literature Overview

Etymologically, the roots of *criticality* are in the Greek term and concept of krisis/κρίσις, which is, according to Koselleck (Koselleck et al. 1982: 619 ff.), a concept within the discipline of premodern medicine. Initially, it was used to describe that specific moment in the course of a disease in which the fate of a patient is determined – survival or death. The term and the underlying notion of a *turning point* later transcended into the fields of theology, law and military.

When we try to understand the potential of the term *crisis*, it is equally important to examine its metaphorical effects. The use of the world *crisis* and *crisis*-related terms evokes images of a radical change of the status quo. As such it became synonymous with terms like *riot*, *conflict* or *revolution* in modern history (Koselleck et al. 1982: 649).

Today, disciplines like mathematics and thermodynamics are using this term in order to describe the likelihood of transformation of certain substances or components between qualitatively different states (Bouchon 2006), hence the connection to the aforementioned *crisis*. Take for example the criticality of radioactive materials, describing the imminence of a chain reaction that cannot be stopped.

In recent years, the concept of the 'tipping point' has been highlighted in technoscience and resilience research to describe these moments (cf. Höhler 2014). In a similar way, Högselius et al. (2013: 265) point out the importance of so-called 'critical events', e.g. incidents or impacts suddenly causing function failures in technical infrastructure systems.

During the early Cold War, the notion of *criticality* was incorporated into the civil defence strategy of the United States. Collier and Lakoff (2008b) show that the term was used in the context of 'vulnerability mapping', e.g. a sort of assessment of vital parts of the country, namely important technical infrastructures. Since then, criticality has been used more and more in political and administrative debates as a prominent label to identify organisations and institutions that are in any way important, relevant or necessary for the continuation of the population's and the economy's supply with goods and services (Bundesministerium des Innern 2009).

The concept of criticality is further known in risk analysis and risk research. In that particular field criticality is used synonymously for 'risk' and 'vitality'. In this context the famous definition of criticality has been developed as a relative measure for the consequences of a disruption to or malfunction of society's security (Bundesamt für Bevölkerungsschutz und Katastrophenhilfe 2017). In practice, criticality or risk analysis combines a critical component and a vulnerability analysis in order to identify the impact of disruptive events on various affected sectors. The impact of disruptive events is divided into different groups: firstly with regard to their geographical dimension and the number of people affected, secondly with regard to the impact on the economy and infrastructural interdependencies, and thirdly with regard to the temporal factors (Theoharidou et al. 2009). In addition, Katina and Hester (2013) are proposing generalisable criteria for criticality: resiliency, interdependency, dependency and infrastructure risk.

Critical infrastructure is an important concept in political discourse. There are numerous national defence or protection strategies for critical infrastructure, e.g. 'The Clinton Administration's Policy on Critical Infrastructure Protection' (The White House 1998) or 'Great Britain's National Security Strategy' (Great Britain Cabinet Office 2010), and other national and international initiatives, like the European Commission (2006). Still, the political concept is related to its etymological roots. If a disruption or breakdown of an infrastructure can cause a systemic or societal crisis, this infrastructure is critical.

Since the turn of the century, critical infrastructure and criticality have been taken up by academic research. The majority of scholars in the field of infrastructure research use the terms criticality and critical descriptively, similar to its use in political discourse. Some of them take as a starting point the services to the population and to the economy which are regarded as indispensable, e.g. the supply of water or energy. Consequently, the technical systems providing these services are labelled as critical infrastructure (cf. Moteff et al. 2003). Other scholars prefer a historical or developmental perspective. They do not start from a set of vital goods and services, but trace how in certain societies or time periods specific services have been considered more and more important and therefore critical as a result of collective preferences (Högselius et al. 2013). Distinguishing between critical and non-critical infrastructures can narrow down the field of research, e.g. as a starting point to investigate interdependencies between them (Rinaldi et al. 2001) – although it might be difficult in some instances to deprive a set of infrastructures of their quality as critical. Like in many political discourses, it is less problematic to incorporate more and more systems into the category of criticality than doing the opposite.

Bouchon (2006: 33-59) provides a comprehensive overview of the different concepts of criticality with respect to infrastructure research and management. In a forthcoming volume to be published by the KRITIS graduate school, the concept of criticality will be elucidated according to its use for practitioners and more theoretical humanities and social science approaches. If applied carefully, criticality can help analyse a multitude of interdependencies in different societies (Engels/Nordmann 2018).

In IT security, vulnerabilities are often classified by severity, with critical vulnerability being the most severe. This classification can be done by using a metric which combines numerical values for the required access to a vulnerable system, e.g. physical or remote, possible interactions of a user, and the control over a compromised system, e.g. complete or partial (First.org 2015). However, this is not the only use of the term critical vulnerability and the etymology of the term is currently an open question.

3 Criticality within Critical Infrastructure Research

Criticality is of course a key concept in critical infrastructure research, as it is an integral part of its very notion. As we have noted above, we do not think that a simply enumerative approach helps us to understand the nature of criticality. Instead, enumerations are helpful to get an impression of which infrastructure is commonly regarded as important. Enumerations help understand the political or economic 'weight' of a given infrastructure. However, the concept of criticality is more than just 'importance'.

Moreover, it is more than the mathematical implications of the extent of impacts and the probability of occurrence, as calculated in contemporary risk analysis. Rather, criticality builds on the attributes of urgency and (public) interest and implies an underlying obligation to act. In practice, it comes down to terms like 'urgency', 'sense of urgency', 'need for action' or 'obligation to act' when talking about critical systems or critical elements.

Thus, criticality is a concept that expresses and helps to analyse relations between different entities. In our case, it is all about the relationship between different infrastructural networks and/or between networks and society. The very nature of a network consists of establishing relations, so that we can argue that the research about criticality of infrastructures is essentially research on the scope and density of interlinking. Simply being part of a network does not constitute a *sufficient* condition for calling it critical. In a given network, an individual node can be more or less interrelated with other nodes, so that the criticality of different components is different. Criticality research is about these differences. Thus, criticality is not an ontological assertion. Infrastructures (or their parts) cannot be critical as such, but only in relation to something that is depending on them. Criticality is a flexible concept that extends beyond binary concepts like 'critical: yes or no'. Rather, criticality can be assessed by degree or in comparison ('more or less critical than …'). Relations are constantly changing, so that criticality is varying in the course of time or in different settings. However, the criticality of an element or a system depends not only on the contexts in a given time frame and a given space. It also depends on the research question or the function ascribed to it (cf. Engels 2018).

Still, there are different uses and understandings of that concept. Since our approach is interdisciplinary, we do not want to prescribe a certain way of understanding, but we aim to highlight certain differences in the current use. We should note at least three major directions, namely (1) deficiency vs. capacity-oriented uses, (2) function-oriented uses and (3) pragmatic uses.

The first direction refers to the difference between deficiency-oriented and capacity-oriented approaches, although these very terms have not yet been used very often to characterise these approaches (Engels 2018). The majority of authors and, above all, the policy and security concepts are starting from the question of failure: what happens if a certain component breaks down or if a service is no longer available due to that? This debate is essentially about society's vulnerability or the weak points of technical infrastructure. It is about deficiency and strategies to cope with that (some of these strategies are elucidated in chapter 6 on preparedness & prevention).

However, several authors have argued not to put too much emphasis on the negative side of failing infrastructure when elucidating what criticality is. Instead, they conceptualise criticality (and critical infrastructure) with respect to its capacities. One example is cited by Bouchon (2006: 37): in the case of a natural disaster, critical infrastructure is not (only) failing infrastructure. Instead, all infrastructure which continues to function and (still) provides vital services that help in coping with the emergency is critical, for example all what is needed to help rescue people, to uphold communication, and so on. Egan (2007: 7-8) points out that the failure of components of one infrastructure can be externalised to other infrastructures. This vision is informed by the approach of preparedness rather than that of protection against functional crises.

Other authors do not even start from the case of emergency, but refer to times of smooth function and perfect service. All infrastructures which enable a society to develop are critical; critical infrastructure should be regarded as a constructive part of modern human life (Fekete 2011: 16; Lenz 2009). This perspective is similar to the idea of empowerment or 'enabling' functions in the theoretical debate about the concept of power (Engels/Schenk 2014): instead of focusing on the perspective of restrictions and constraints it puts forward constructive effects and uses for the criticality concept.

The capacity-oriented approach is important because it helps avoid a restriction of critical infrastructure research to questions such as security, given the fact that the 'securitisation' (Buzan et al. 1998) of many policy fields in recent decades, i.e. the framing of many issues as problems of (in)security, instead of (in)justice for example, has not only had positive effects. Moreover, several authors (e.g. Aradau 2010; Collier/Lakoff 2008a) blame the securitisation approach in critical infrastructure research for focusing on a certain class of problems, for instance the abovementioned sudden function failures causing cascading failure effects.

Andreas Folkers (2012) emphasises the political and ideological impact of criticality as a key concept in public life. In his analysis, the criticality/security approaches to infrastructure protection are inseparable since criticality is a concept developed by the military during the Cold War era and later put forward by politicians who wanted to counteract the dangers of terrorism and violent threats. Therefore, the very idea of critical infrastructure in public and academic discourse has centred around the concept of fear. Moreover, the concept of criticality allows decision makers to take choices and to establish hierarchies between more or less critical components or networks, giving priority to the former's protection. As a result, the egalitarian idea of the networked society has been given up, and certain nodes are much better protected than peripheral regions.

The second direction refers to those approaches that are oriented towards the function or functionality of infrastructures or infrastructural components. Again, this specificity has not yet been elucidated in infrastructure research. At the heart of function-oriented approaches lays the assumption that function or functionality is the result of an interplay of individual components. Initially it does not matter if these components are technical, social, political or whatever else. Depending on how the overall function or functionality is defined, individual components can be assessed as being critical if their failure compromises the function of the whole or of other components that depend on them. The connection to crisis is apparent. The failure of these critical components triggers the abovementioned 'sense of urgency' and urges the reestablishment of functionality or function. In function-oriented approaches, criticality is assessed in relation to an overall concept of a given or desired function or functionality.

If, for instance, running is the main function of a train, then tracks are a necessary condition for this function. They are therefore *critical* for running a train. While many components are critical for a system, some components can be more critical than others. The *criticality* of a component is high if others are heavily reliant on its proper function. *Criticality* is lower if those components play no major role for the perpetuation of the system's function. According to function-oriented approaches, the heart is more critical for the function of the body than an individual kidney is, given that the function of the body is defined as *perpetuating life* and not *upholding the highest diuretic function*.

A large number of infrastructure research theories are variations of function-oriented approaches. In this context, we often find texts highlighting a difference between 'systems-based' and 'consequence-based' concepts of criticality (Egan 2007; Metzger 2004a). Still they have in common that criticality is assessed in relation to a given or desired function or functionality.

Systems-based criticality assesses the relevance of technical components for a system, or the importance of subsystems for overall systems. In many cases, systems-based criticality is more focused on technology. One of the classical applications of systems-based criticality is the phenomenon of cascading breakdown effects in interrelated networked infrastructures (cf. Little 2009) (e.g. IT infrastructure and energy infrastructure are critical for the running of communication infrastructure, communication infrastructure is critical for the running of transport infrastructure, etc.). The consequence-based approach does not exclusively look at technological systems. In fact, it is focused on society and population: it describes possible harm to the people as a result of infrastructure failure, as reflected by the abovementioned definition of the Bundesministerium des Innern (2009) and all the other protection policy concepts.

The difference between these two approaches is based upon their different perspectives: systems-based criticality 'looks' from the bottom level of the single component onto the complex interrelatedness and interdependencies of material arrangements and systems. The consequence-based approach starts from a 'big picture' of the services expected by the society. It looks down from the top of social facts so to say, and identifies the services provided by technical systems and the possible effects of their loss. From a societal point of view, the difference between systems- and consequence-based criticality is about differences between the world views of providers or users.

However, it is important to note that these are only different perspectives, for two reasons. First, there is no strict separation between the material world and technology on the one hand and the social world on the other. Namely we think the concept of 'system' does not exclude the social world but helps us to conceive of the interrelatedness between artefacts and social phenomena. Second, *both* fall into the category of function-oriented research, provided that they are interested in the consequences of failure, be it the consequences for the water supply system itself, or the resulting hygiene practices in the society affected by that failure. In the latter case, the hygiene practices (teeth brushing, use of water closets, other forms of body care) are functionally dependent on the provision of fresh water. If that provision fails, the practices are seriously harmed. With respect to the function of the whole set of devices and practices, everything is connected to the overall goal of disease control. Personal hygiene plays an important role in disease control – it is critical for it – and freshwater supply is critical for a certain standard of personal hygiene. Technological and social devices, components and practices are all connected on a functional basis.

The third direction to conceive criticality can be termed the pragmatic approach. This concept refers to criticality as being ascribed within a discourse rather than assessed according to a function. Pragmatic approaches focus on the metalevel of how criticality is constructed by experts, analysts, politicians, practitioners or scholars. In order to understand the variations of pragmatic approaches it has to be clear that they are not concerned with the practicability of an ascription of criticality. They are rather concerned with the ascription of criticality through pragmatic acts, speech acts as it were, rooted in the conventions of our language. Pragmatic approaches are however related to function-oriented approaches, because they also build upon the idea that some components become critical for the function of a system. Nevertheless, it cannot be stressed enough that both function-oriented and pragmatic approaches have to be distinguished in order to make clear statements about the criticality of certain elements.

Pragmatic approaches analyse how political, scientific or social discourses take advantage of this relative criticality in order to proclaim an absolute or ontological criticality. This criticalisation[1] works very similar to the theory of securitisation in security studies (Balzacq 2005: 171 ff.). Aradau (2010) has pointed out the consequences of the securitisation of critical infrastructure protection, coining 'materialisation' a certain idealised vision of infrastructure which tends to blank out the fact that technical systems are regularly prone to failures and disturbances.

Pragmatic approaches to criticality are about the complex interplay of actors, context and audience. Through recourses to functional criticality certain elements are assumed to be critical within the discourse. Criticalised components do not become critical in relation to other elements in accordance with a function, but are proclaimed as being critical. This generates a sense of urgency which can be utilised to justify political action or scientific research. Mentioning criticality might also be used within a discourse to appeal to the reason of the audience. Difficult connections might be presented in very easy narratives that seem to have a clear solution; for example, it might seem like common sense to secure those elements that are so clearly vital for a society.

---

1 Special thanks to political scientists Markus Lederer, Judith Kreuter and Benedikt Franz from Technische Universität Darmstadt for the fruitful discussion on this topic.

For critical infrastructure research, this poses some difficult challenges. Criticalisation conflates technical relationships and social contexts. The functional criticality of infrastructures is reintegrated into political, scientific and social discourses, thus transforming them from *matters of fact* into *matters of concern* (Latour 2004: 231 ff.). As such, they become part of a negotiation process. On the one hand, *making* certain infrastructures critical puts them on a par with other critical infrastructures. On the other hand, this labelling can cause problems, e.g. if the infrastructures that are labelled in this manner are of significantly lower functional criticality than they appear within the discourse.

The great advantage of interdisciplinary infrastructure research is that it enables us to combine function-oriented and pragmatic approaches. Normally, humanities and social sciences take up the part of the pragmatic approach, whereas the engineering perspective focuses on the function side. As we have seen, both are interrelated, at least in modern societies, insofar as the ascription of criticality is only convincing if it is based on the engineer's expertise attesting the functional role of the components in question. Quite the same might be true the other way round: if an engineering expert wants to assess the criticality of a certain component or system, he is depending on the rules of the political or societal discourse. It is not a coincidence that the highest degree of criticality is always ascribed to systems which endanger human lives when failing (cf. Egan 2007: 12-13). Interdisciplinary research on critical infrastructure consists of reflecting this interrelatedness without denying the methods and goals of every single discipline.

4 Open Questions and Problems

The problem with criticality is the variability of its content – as we have seen. Mainly we have distinguished the opposition of deficiency and capacity as well as the variations of function-oriented and pragmatic approaches as being expressed in this concept. There is no need to merge these conflicting perspectives and variations into one single use of the term. Rather, scholars should be very well informed about the differing dimensions and should constantly reflect upon their own position and approach.

The notion of *criticality* is still a very ambivalent one. It can be utilised to point out the most important and most vital elements of technical, socio-technical, political and social relationships. It can also be used to define priorities in resource spending. If, however, *criticality* becomes part of a specific discourse, the lines between the function-oriented and pragmatic approaches can easily become obscured. Participants in such a mixed discourse can easily commit a category error and interpret the usage of the term incorrectly. In that case it is important to analyse how a specific notion of *criticality* was derived. Otherwise one may fall victim to certain dynamics within a discourse or to the many preconditions of supposed functionality. Therefore, *criticality* should be an object of constant reflection. Any serious research should clarify where the *criticality* of its infrastructures is coming from.

Another important problem is to determine the boundary between 'critical' and 'not yet critical'. Within the function-based approach, it is comparatively easy to ascribe the feature of criticality to a given system or component, and it may also not be too complicated to rank two or more components of a given system according to their degree of criticality within the system. As criticality is a relational category, it is however hardly possible to identify an exact threshold level between criticality and non-criticality (Fekete 2011: 19). Therefore, it is very difficult to measure criticality exactly. Nonetheless, practitioners will be continuing to demand approaches in order to classify elements, subsystems or systems according to their criticality.

Vulnerability

Stephanie Eifert, Alice Knauf, Nadja Thiessen

The concepts of vulnerability and resilience are often used complementarily. Both of them focus on the relations of systems towards stress or perturbations. Whereas resilience concentrates on the possible resources and capacities that would help the system to maintain or achieve a desired state, vulnerability usually focuses on the flaws and weaknesses of the system. However, although this vision is popular, there are also strong arguments against a dualistic view of the two concepts. Therefore, we elucidate the concept of vulnerability in this chapter before subsequently turning to resilience.

1 Ad-Hoc Description

W. Neil Adger (2006) describes the 'lowest common denominator' of vulnerability as follows:

- the stress a system is exposed to
- the system's sensitivity to stress
- its adaptive capacity

This definition relates the *stress* to a *system* and its *exposure* as well as its (low) *capacity* to adapt.

However, there are two important shortcomings of this definition. First, it puts the focus on 'external' (Chambers 1989; Bankoff/Frerks/Hilhorst 2004) or 'hazard-dependent' (Bouchon 2006) types of vulnerability. It neglects more or less the inherent or internal vulnerability factors of a system. Infrastructure research, however, has tried in recent decades to recognise and to describe the systems-inherent problems and vulnerabilities, expressed in concepts like 'normal accidents' (Perrow 1984) or 'cascading failure' (Bak/Paczuski 1995; Little 2009). The complexity of the cross-linking, mutual dependency etc. of technical infrastructure must be considered as a potential factor of vulnerability in our research context.

Second, due to its technology-style language this description tends to neglect social or cultural factors. Adger's definition should be complemented by what Steve Engler calls 'social vulnerability'. That includes 'defenselessness, meaning a lack of means to cope without damaging loss' (Engler 2012: 159). It applies to human individuals as well as to social groups or even societies (Collet 2012; Birkmann 2011; Brklacich/Bohle 2006).

2 Literature Overview

As with almost all our key concepts, the concept of vulnerability is used in various fields of research. Each discipline defines the term in a different way. Even within single disciplines, we still find striking differences with respect to the content, and very different opinions regarding the usefulness of this concept (cf. Adger 2006; Christmann et al. 2011; Bürkner 2010; Birkmann 2011). However, vulnerability has been used since the 1990s as a category to express or to analyse the susceptibility of ecological, social, economic and technological systems. Christmann/Ibert (2012) and Bürkner (2010) give useful accounts of the term's various applications. The following paragraphs will present different research perspectives and their individual uses or interpretations.

2.1 Different Research Traditions

The Latin roots of the term are *vulnerare* and *vulnus*, which can be translated as 'wound'. Thus, in medicine and psychology, vulnerability is a concept to describe the disposition, genetic or acquired, or diathesis of a person to be affected by a certain disease or an illness. Vulnerability here is conceived in an individual perspective. With the help of experts, the person in question can recognise and comprehend his or her own vulnerability. By changing behaviour, the person can even reduce his or her likelihood of falling ill. Interestingly, there is no external factor or event causing vulnerability (which shows once more that Adger's aforementioned ad-hoc definition lacks certain important features). Psychologists prefer the term 'cognitive vulnerability'. Some authors underline the significance of this concept in the *Encyclopedia of Cognitive Behavior Therapy* (Alloy/Abramson 2005; Riskind/Black 2005).

In engineering science, vulnerability is a very prominent concept. Often, it is conceived as a possible security gap. First, vulnerability is hard to detect in complex networked systems. A single component's failure can put at risk the security of a complete technical system or subsystem. Second, there are different possible causes. There may be external risks, like attacks by criminals or terrorists, or the impact of natural hazards may cause failures. There is however another possibility: namely vulnerability may be induced by the very system, e.g. as a result of its complexity (Perrow 1984), without the occurrence of external factors (Kröger/Zio 2011). In particular, interdependencies between individual components can cause cascading effects (Rinaldi/Peerenboom/Kelly 2001).

In spatial science, human geography and planning vulnerability is conceived as susceptibility of the human–environment system. Identified risks are often natural hazards and ecological changes. However, these research perspectives are not primarily focused on risks and threats, but rather on vulnerability measurement, analysis and mapping. These approaches promise insights that may help to improve the resilience of the system. This approach is somewhat similar to the use of vulnerability in the medical and psychology context – although it does not favour the internal perspective. However, there are frequent references to 'social vulnerability' (Fekete 2013; Adger 2006).

In human ecology, vulnerability is conceived as potential or actual impairment of social systems or individuals. Development research has been working with the concept of vulnerability for a long time. Vulnerability means in this case the precarious access of people to essential goods and services. Yet this approach excludes the individual perspective like physicians have, because vulnerability here is conceived as a structural phenomenon which depends largely on economic, social and political conditions and capacities. When structural vulnerabilities meet a critical event, a social catastrophe is likely to occur (Chambers 1989; Watts/Bohle 1993; Wisner et al. 2003).

2.2 Need to Consider Social and Cultural Factors

Regrettably, a whole class of studies neglect the social and cultural factors contributing to a system's vulnerability. There are many examples of 'vulnerability mapping' or 'vulnerability analysis' in infrastructure research which omit elements like the role of stakeholders, society or everyday culture (cf. Bouchon 2006; Birkmann 2011). However, many scholars have picked up the 'social question'. Wisner et al. (2003) have suggested taking into account the living conditions of people in order to assess their vulnerability. Other researchers not only demand the inclusion of social contexts when considering vulnerability (Adger 2006), but propose always conceptualising vulnerability in a 'holistic' manner, including physical as well as social conditions (Engler 2012). More sophisticated approaches demand the differentiation between emic and etic (culturally internal and external) research perspectives (Schenk 2015: 12; Collet 2012: 19).

We think these demands are justified – for the simple reason that, for example, wealthy societies have a different sensitivity to stress than poor societies. An example concerning economic policy may be the reliability of German railway services. The (part) privatisation of Deutsche Bahn since the mid-1990s made cost reduction imperative. As a result, some lines were closed, the maintenance of signal systems and turnouts was neglected, personnel was reduced – in short: redundancy has been diminished. As a result, minor disruptions (like flu outbreaks among staff or points failures) have led to considerable delays across large parts of the network.

Yet it is not all about economic decisions and technical standards. The 'paradox of vulnerability' ('Verletzlichkeitsparadoxon', cf. Folkers 2012) teaches us how important societal contexts and culturally driven practices are. It works as follows: the more a society has built smoothly functioning infrastructure capacities, the more it is disburdened from coping with functional disruption in its day-to-day practices, but the more it is vulnerable to a single breakdown event. The better the reliability of the systems, the rarer a society is confronted with service blackout, and the higher are the costs for coping with it. In a society facing energy blackouts every day, people develop strategies like buying power generators or candles, so that they include these blackouts into their daily routines. In societies without daily blackouts, there is no incentive to develop such strategies of redundancy. In the case of a long-term power outage in developed societies, people sit in the dark, comestibles rot in the refrigerators, etc. In this context, Högselius et al. (2013: 10) propose the term 'user vulnerabilities'.

These examples should have made clear that in order to understand vulnerability in its complexity, it is essential to consider a plethora of social, economic, cultural and political frameworks and conditions.

2.3 Major Differences

As shown, the objects of research are varying, depending on the disciplines and perspectives. Vulnerability research may focus on individuals, groups or systems with social, ecological, economic or technical methods. There are various risks in focus, such as natural hazards, climate change, economic or political crises, poverty, technical failures, criminality or terrorism.

We have detected four major differences in these approaches: 1. Internal vs. external: Some take into account external *and* internal risks to a system or an entity at the same time, some instead concentrate on one of them. 2. Event vs. structure: Some are concentrated on single shocks or events, others are more preoccupied by structural and long-term weaknesses. 3. Technology vs. society: Some are more interested in technical failures, whereas others are concerned with social behaviour or 'social vulnerability'. 4. Components vs. whole systems:

Some approaches are interested in understanding vulnerability starting from individual components, others prefer a comprehensive, or holistic, comprehension of vulnerability.

For our purpose, namely interdisciplinary critical infrastructure research, none of these perspectives should be excluded. We should rather combine them in order to benefit from a multitude of insights, according to our specific research problem (pledges for considering social *and* technical, internal *and* external factors etc.; cf. for example Fekete/Tzavella/Baumhauer 2016; Graham/Debucquoy/Anguelovski 2016; Pescaroli/Alexander 2016).

2.4 Negative or Positive?

In general, vulnerability is conceived in a perspective of weakness – a high degree of vulnerability is equivalent to high exposition and sensitivity, and a low level of adaptation with respect to stress. However, the negative implications of the term have been questioned by some authors. These scholars are influenced by fluid models of society and/or technical systems. They point out the potential of learning, optimisation and transformation as a result of vulnerabilities (Bijker/Hommels/ Mesman 2014: 22; Gallopín 2006: 295). They contend that a certain degree of vulnerability may be very beneficial, if not necessary, for a social system as a prerequisite to remain open for change and to reduce path dependency. This debate on the positive side of vulnerability may be paralleled with the abovementioned approaches to considering criticality in a positive way.

Nevertheless, there is a logical problem remaining: as we have seen, vulnerability is closely linked to the concepts of stress and risk, which are inherently disruptive, if not destructive, forces. Therefore, the 'positive' side of vulnerability may be considered as yielding 'awareness' or attentiveness in a society, resulting from the knowledge about certain vulnerabilities, which motivates a process of continuous optimisation or adaptive learning. This would be closely linked to the concept of preparedness. Yet there is an inherent contradiction to our ad-hoc definition: vulnerability is partly *equivalent to poor adaptation capacities.* So we cannot say that vulnerability causes adaptation, but improving adaptation strategies will reduce vulnerability.

Maybe it is helpful to consider this argument in a more dialectical way, comparable to the theory of 'creative destruction' / 'Schöpferische Zerstörung' (Schumpeter 2005 on capitalism). In this framework, minor disturbances or failures, due to vulnerability, may give a society the opportunity to react and to optimise, resulting in lower vulnerability. By definition, vulnerability itself cannot be considered positive, because generally the reduction of vulnerability is the desired result.

3 Vulnerability within Critical Infrastructure Research

Vulnerability as a research concept offers many opportunities for interdisciplinary research, e.g. in the fields of engineering, political, economic, social and historical sciences and urban planning (cf. Collet 2012). As we have explained above, critical infrastructure research absolutely must take into account the technical aspects of vulnerability as well as the social or human dimension ('user vulnerability', 'social vulnerability'). Dominik Collet sharply criticises that many studies do not consider culture-specific differences when analysing vulnerability in infrastructures (Collet 2012). We contend that even the very same networked infrastructure has a different type of vulnerability according to the society it is run by.

For instance, we have to look closely at power relations, for example entanglements between social groups and political or economic stakeholders, if we want to assess factors of vulnerability. In particular, vulnerability often cannot be assessed realistically without taking into account long-term processes. The vulnerability of the rail network depends on the age of its components, including in Germany more than a few bridges built in the interwar period prone to processes of decay induced by corrosion, humidity, etc. Vulnerability research should thus be open to long-term investigations, considering the historical perspective and taking into account the timely development of vulnerability factors. Crises in networked infrastructure often result from human-made problems and path dependencies. Hence the identification of technical and social aspects in space and time causing vulnerability is necessary to better understand them; a holistic approach is necessary. These reflections may also help in understanding why recent approaches to vulnerability often stress the processual character (of the emergence) of vulnerability (UNISDR 2004; Endreß/Rampp 2014; Christmann et al. 2011).

Strategically spoken, vulnerability analyses provide a chance to reveal the interrelatedness of technical and social factors and to criticise – with good arguments – the lack of awareness of human factors in many studies (Bürkner 2010). Vulnerability studies on infrastructures thus provide the opportunity to link the perspective of engineers, who aim at improving technological solutions, and of the humanities, which tries to learn from experiences of the past (cf. Bijker/Hommels/Mesman 2014). In other words: the vulnerability concept opens up perspectives for inter- and transdisciplinary learning processes (cf. Wisner et al. 1977).

We think there can be no standardised method applicable to every infrastructure. So we do not propose a specific model for the vulnerability analysis of infrastructure. However, we contend that analysing vulnerability (of a technical or a social entity, or both) requires the careful consideration of at least three aspects. We think this is a prerequisite for transparent and state-of-the-art research on vulnerability. This applies to all disciplines, regardless of their specific methods. Our position is strongly informed by, but not totally identical to, Christmann et al. (2011, esp. pp. 25-26).

First, vulnerability is object-related. So it is indispensable to first define the object of the study very precisely. What is the entity of which we want to assess vulnerable features? In critical infrastructure research, this may be a technical component like a transformer station, it may be a networked system like the electricity grid in a certain region, or it may be several interrelated systems. Moreover, we may focus on certain functions or services provided by infrastructure, e.g. analysing the vulnerability of the energy supply. Again, we have to specify if we adopt the perspective of users or 'the city' as a whole (cf. consequence-based approach in the chapter on criticality), or if we focus on the technical interplay of components and sub-networks (cf. systems-based approach). This list is far from being comprehensive – it gives us an impression of how diverse possible objects of vulnerability research can be.

Second, vulnerability is risk-specific. This argument is very close to the first point. Not only does our object of research has to be specified, but we also have to make clear which kind of risks we consider significant in the context of our vulnerability assessment. It is not reasonable to analyse for example the vulnerability of a given urban population to the risk of an energy blackout and to the risk of climate change in one study based on the same parameters. The same applies to the risk of flooding and the risk of earthquakes, when we try to assess the vulnerability of public transport in a town located on a river. Different risks have to be researched separately. As a consequence, however, we will not be able to assess the overall vulnerability of a given technical infrastructure system, but only with respect to selectively chosen issues.

Third, vulnerability research is reliant upon parameters. For every single class of objects and risks, specific parameters have to be fixed. They have to be derived from a preconceived (e.g. theory-based) model. The model cannot be defined once and for all. Rather, the choices mentioned above (which object?, which class of risks?) form the basis of a tailor-made model for every specific case study. In the context of our STS-based research, however, we will give preference to parameters which take into account the technology–society permeation.

To sum up: we think vulnerability assessment is highly dependent upon different variables chosen by the scholar involved with this research. Vulnerability is not a given feature of a certain infrastructure; rather it is a perspective that helps us organise research into possible weaknesses in a system. Vulnerability, taken as a perspective or as a heuristic tool, is never fixed. Therefore we argue that vulnerability is not a property of an infrastructure, but a process. Or, to be more explicit: academic vulnerability assessment is a process. This position could be called relativistic or constructivist.

4 Open Questions and Problems

This section deals with three of the most pressing challenges, and their linkage to the context of critical infrastructures. These challenges are: the actual measuring of vulnerability, its negative connotation and the integration of trade-offs in vulnerability thinking and governance.

W. Neil Adger pointed out the challenge of measuring vulnerability more than ten years ago already (2006: 274), and to this day it has not lost much of its relevance. In the (political) debates on the vulnerability of critical infrastructures, in most cases the focus is on people's vulnerability due to the reliance on the services provided by the infrastructure (cf. the abovementioned 'paradox of vulnerability'). Conceived in this way, vulnerability is a social problem or a public one, very closely linked to the assessment of consequence-based criticality. Moreover, the expectations concerning services provided by infrastructure differ from society to society, sometimes even from individual to individual. Thus, vulnerability becomes a matter of subjectivity. Of course, this does not simplify the task of measuring vulnerability when it goes beyond single cases. To the present day, we have not found a manageable and intellectually convincing approach to measure vulnerability across different societies and across time and space. Nevertheless, it would be very beneficial if we had such a tool.

A second challenge arises from the negative connotation of the term 'vulnerability'. We have seen that there are approaches to consider positive aspects of vulnerability. However, it is politically problematic to consider weaknesses as a kind of strength. It would hardly be acceptable to the public to conserve intentionally certain vulnerabilities in energy provision or railway systems. Deliberately running the risk of provision or service breakdown would kill the legitimacy of any responsible organisation. However, we know perfectly well that no system is free from vulnerabilities. So we should debate under which conditions is it useful to accept a certain degree of vulnerability in favour of openness? Which aspects of an infrastructure's vulnerability can open up ways for adaptive learning?

Last but not least, modern cities face a broad range of challenges. Alongside reducing the vulnerability of their critical infrastructure systems and societies, topics like climate change mitigation and adaption, reducing poverty as well as social division and exclusion are at the top of the agenda. Measures applied in the context of each of these topics can have effects in different dimensions, for example in the social, ecological, economic or technical spheres. On top of that, these themes are highly interlinked. Considering trade-offs between different spheres and different policy areas in a broader scale is then a prerequisite for proficient scientific work and governance. Lorenzo Chelleri et al. (2015) provide valuable insights concerning the concept of resilience, indicating that it is possible to build up a critical and manageable framework which grasps trade-offs on multiple scales and at different times. For the concept of vulnerability, comparable frameworks for critical infrastructure protection are still needed.

Resilience

Ivonne Elsner, Andreas Huck, Manas Marathe

As mentioned above, the concepts of vulnerability and resilience are somehow complementary, both focusing on systems exposed to stress or perturbations (Miller et al. 2010). Whereas vulnerability usually highlights weaknesses, the concept of resilience concentrates on the abilities of the systems to resist, recover or adapt.

1 Ad-Hoc Description

Resilience is the capacity of a system to absorb and cope with perturbations. There are two major conceptions of resilience. First, resilience describes a system 'bouncing back' to its original state after a shock – this is equivalent to 'recovering'. Second, resilience describes a system 'bouncing forward' to another state in the case of a perturbation – i.e. 'adaptation'.

This would mean the system demonstrating either one or a combination of the following capacities if confronted with a perturbation:

- *absorbing* (preventing a crisis)
- *recovering* (short-term coping and 'bouncing back' to the original state)
- *adapting* (long-term coping and 'bouncing forward' to a new state)

A definition of resilience as used in disaster risk reduction is '[t]he ability of a system, community or society exposed to hazards to resist, absorb, accommodate to and recover from the effects of a hazard in a timely and efficient manner, including through the preservation of its essential basic structures and functions' (UNISDR 2009: 24).

2 Literature Overview

The Handbook of International Resilience recently edited by David Chandler and Jon Coaffee (2017) shows the broad variety of disciplines using the resilience concept. It includes contributions from the fields of political sciences, philosophy,

© Springer Fachmedien Wiesbaden GmbH, part of Springer Nature 2018
J. I. Engels (Ed.), *Key Concepts for Critical Infrastructure Research*,
https://doi.org/10.1007/978-3-658-22920-7_4

anthropology, sociology, geography, environmental and sustainability sciences, disaster studies, civil engineering, resource management and urban studies. The concept of resilience is also used in biological ecology, psychology and medical sciences. This wide range of disciplines not only have different understandings of the term resilience, but also employ different methodologies when using the concept. However, it is beyond the scope of this paper to touch on the use of the resilience concept in all these different disciplines. Resilience is even used in political and popular discourse. States try to become 'resilient'. Guidebooks and self-help literature are presenting ways for individuals to 'become resilient' when they face life crises. Of course, we are not dealing with that kind of literature. However, every researcher using a concept as popular as this should bear in mind that there are parallel discussions blurring the meaning of that concept even further.

As our work is dedicated to critical infrastructures in cities, this section focuses on the context of urban and infrastructure management and planning.

2.1 Origin as a Scientific Concept

Despite their differences, contemporary understandings of resilience within scientific discourses can be traced back to the same root. The concept of resilience was taken up by ecological scientists along with the rise of systems thinking in the 1960s. Especially the US–Canadian ecologist Crawford Stanley Holling not only shaped but also contributed massively to the upturn of the concept. He describes resilience in his paper entitled 'Resilience and Stability of Ecological Systems' as '[…] a measure of the persistence of systems and of their ability to absorb change and disturbance and still maintain the same relationships between populations or state variables' (Holling 1973: 14). Holling considers engineering resilience to be a technical or physical system's capacity to bounce back to its original state after external shocks (single equilibrium approach), and ecological resilience as an ecological system's ability to persist and to adapt to changing circumstances with a multi-equilibrium approach. Hempel & Lorenz (2014: 30) argue that the career of resilience was the result of an overall 'ecologization' of thought since the 1970s, replacing static or teleological representations of society by dynamic awareness. From this basis, the concept was further developed. For example, Davoudi et al. (2012) go beyond an equilibrium to a more evolutionary approach to resilience by understanding systems as constantly changing over time (with or without external pressure). In this perspective, resilience is conceived as the ability of complex, non-linear and self-organising socio-ecological systems to change, adapt and transform in response to stresses and strains. Whilst the concept of vulnerability mostly focuses on known threats, resilience offers to include yet unknown risks and the ability of a system to cope with them. In this sense, resilience can be read as an answer to a growing sense of uncertainty in a state of constant, yet diffusing, risk (Beck 1992).

2.2 Resilience as a Postmodern Approach

Like every scholarly concept, resilience is a result of certain historical conditions in its time period. A few lines above we hinted at the fact that resilience is political, and that it is perfectly comparable to neoliberal economics and governance approaches. We also made clear in the respective chapter that criticality's career in recent politics and protection strategies has much to do with the reactions to 9/11 and the terrorist threat, as well as the general tendency of securitisation during the last decades. In order to understand the career of resilience, however, we must go beyond the level of politics – rather it is worth taking a look at postmodern world views or systems of ordering the world (in this section we are following Höhler 2014, Aradau 2014).

In classical modern and industrial thought, possible threats or risks have been externalised – technical systems were conceived either on the principle of robustness or of prevention. They have been conceptualised as being inherently statical, or, if not, being enhanced in a way that they became stable. With the emergence of cybernetics and the rise of risk-oriented thought from the 1970s onwards, the conception of technical (and ecological!) systems changed: from then on, they have been considered as being dynamic. Moreover, failure was considered to be an inherent or unavoidable part of the system. The loss of control, which had been extremely undesirable in modern times, is now accepted as a given condition.

The word resilience, which had been used in materials science since the 19th century to describe the 'elasticity' of certain physical materials, has now become a new scientific paradigm, due to its metaphorical power. Transforming a description (of ecosystems) into normative prescription, technical systems (and even societies) now have to be thought of as ever-changing, adaptive, smart, transforming and, ideally, as 'bouncing'. The latter became the central metaphor in resilience research. Working with the bouncing and elasticity metaphors is thus deeply informed by postmodern styles of thought, systems of ordering and practices (in governance or in the economy, for example). This explains the immense popularity of the concept in different academic fields.

2.3 Urban Resilience

Urban scholars and practitioners entered the resilience debate in the 1990s (Coaffee et al. 2015: 250) from different orientations (Pizzo 2015: 134). For that reason, urban resilience debates range from disaster risk management (e.g.

Matyas/Pelling 2015) and secure urbanism (e.g. Coaffee 2009) to urban governance (e.g. Duijnhoven et al. 2016), planning (e.g. Goldstein 2012) and design (e.g. Desouza/Flanery 2013). Quite often, these debates centre around issues of climate change and extreme weather events (Hodson/Marvin 2010; Jabareen 2015), documenting the 'ecological tradition' of the concept. Moreover, discussions start to emerge on urban ecosystem services (e.g. McPhearson et al. 2015) as well as on the potential failure of infrastructures in urban areas (e.g. Graham 2010). As Johnson and Blackburn (2014) argue: 'Each city sees different aspects as important for resilience, and these are context specific.' In this sense, urban resilience can also refer to different urban development processes and different time dimensions: recovery (short-term) – adaptation (mid-term) – transformation (long-term) (Chelleri 2012). Generally, however, elements of urban resilience are often described by characteristics like awareness, diversity, integration, self-regulation and adaptability (Rodin 2014).

2.4 Criticism of Hidden Agendas in the Concept

Resilience is often presented as a politically neutral approach (Pizzo 2015: 134), e.g. in the sense of an analytical tool in the context of climate change for assessing urban physical structures, functions and services (Olazabal et al. 2012: 11). However, at the same time it is increasingly exploited in a normative sense by stating that '[…] enhancing adaptive capacity should be the overall goal of resilience […]' (Klein et al. 2003: 43). Hence, the concept becomes inherently political (Davoudi et al. 2017), even if not used by politicians. This has been noted by several authors. Many of these critical accounts of resilience contend that resilience fits very well into neoliberal discourse on self-optimisation, responsibilisation and governmentality (Chandler 2014; Höhler 2014). Main points of criticism are its negligence of political and economic interest and power, its normativity and its (mis)use as a tool for neoliberal governance (Joseph 2013; Swyngedouw 2009; Zebrowski 2013). Moreover, it has been argued that resilience is in its essence just another political buzzword (Joseph 2013; White/O'Hare 2014). Hence, critical scholars question the political neutrality of the concept and the neoliberal agenda behind it.

2.5 Critical Infrastructure Resilience and Disaster Risk Management

Within the context of critical infrastructures, the resilience concept originated from debates on infrastructure protection in the field of security research in the US during the Cold War period. In this context, the research on the enemy's dependencies on critical services and their vulnerabilities in service provision networks led to the recognition of America's own dependences and liabilities (Collier/Lakoff 2008b; Lakoff/Collier 2010). These debates evidently intensified over the course of the last decades (Brassett/Vaughan-Williams 2015: 40). They have been spurred by events like the 9/11 terror attacks in the USA, Hurricane Katrina in New Orleans and the Japanese earthquakes and tsunami leading to the meltdown in the Fukushima nuclear power plant. In this way, the debate on critical infrastructure resilience has been widened from a mere matter of security research to one of public policy, civil and technical engineering, urban planning and design, and disaster risk management. In disaster risk management, resilience currently gains prominence as a concept that stresses the notion of preparedness & prevention (Collier/Lakoff 2008b; Medd et al. 2005).

Ways in which the resilience concept is approached analytically and practically differ as much as the multiple disciplines making use of it. Various interdisciplinary approaches use different qualitative and quantitative indicator-based methods to measure urban resilience (UN-Habitat 2017) or the resilience of infrastructure systems (Prior 2015). However, it should be stressed that such attempts mostly provide a comparative analysis rather than an absolute, objective resilience value. Therefore, there is a lack of common agreement on how to measure resilience and if this is even possible or desirable. In this respect, the state of resilience assessment is comparable to vulnerability research.

3 Significance for our Research

The resilience concept has entered urban and infrastructure debates during the last decades. Against the background of poorly functioning or unwanted systems, it can be questioned whether resilience in a conservative understanding is always desirable (Vugrin et al. 2010: 11). Because of its broad usage and plurality of understandings, definitions and methods and the resulting complexity of the concept it is not easy to find a common understanding across disciplines and research areas. So, resilience is often used as a buzzword in political reports as well as in research agendas and is, therefore, lacking conceptual exactness. Thus, from an analytical point of view, it is of utmost importance to look at the concept and its probable impact on the acquisition and allocation of resources. On the other hand, the popularity of the concept enables us to link the possible findings in infrastructure resilience to other debates in the political realm.

Considering resilience in its different facets as described in the paragraphs above, this concept offers different benefits for infrastructure research. Researching resilience takes into account the fact that technical as well as social systems are never static. It reflects the process character of both (Endreß 2015), and it enables the researcher to conceptualise the processes that bind together technical artefacts and human action. Resilience acknowledges the character of critical infrastructures as complex, adaptive and socio-technical systems (George Mason University, School of Law 2007). Moreover, the concept of resilience seems to be suitable for dealing with critical infrastructure interdependencies and potential cascading effects (Coaffee/Clarke 2016: 2). The implied understanding of resilience goes beyond a sectoral, purely equilibrium, approach and recognises that it is both human and technical factors which determine the extent and effects of a crisis (Bach et al. 2013; Rodin 2014). Therefore, notions of social and environmental resilience enrich the previously limited understanding of engineering resilience related to urban infrastructures as complex, adaptive and socio-technical systems (Sikula et al. 2015). The concept of resilience offers the benefit of considering the notion of change as an inherent part of the system, as implied in the bouncing metaphor. This allows it to be used across numerous spatial and temporal scales.

There is one major difference in the use of resilience which often refers to two schools of thought – namely the abovementioned (cf. ad-hoc description) difference of 'bouncing back' versus 'bouncing forward'. Both can be very fruitful for infrastructure research, not least because their metaphorical force is big enough to facilitate interdisciplinary understanding. In fact, they are describing different, or even opposite, philosophies of how systems, societies, technologies and people should react to shocks or problems – the former preaching a kind of conservative world view, the latter supporting the idea of evolution and adaptation. Again, as with all the other concepts elucidated here, it depends very much on the implicit premises of our research how we can use these different approaches. With respect to resilience, however, it is crucial to make very clear on which side of the divide the researcher positions himself – 'bouncing back' or 'forward'. In the first case he or she ascribes to the system in question (infrastructure, society or both) a desirable state before the perturbation; in the second case, the initial state is considered deficient.

However, both positions have serious shortcomings. When opting for 'bouncing back', we must explain if and how a way back is possible, because every process of bouncing necessarily entails adaptations, changes, if not even transformations of the given system. You never step into the same river twice, as a proverb says. Put simply, it is a matter of time: history does not repeat itself, so you cannot really come back to the initial state. To solve this problem, of course, we can single out a system's specific function to be restored (cf. function-oriented approach to criticality). When opting for the 'bouncing forward' approach, we are confronted with yet another problem. This problem is about the identity of the system under scrutiny: when adaptation and profound change are the very features of bouncing forward resilience, then we have to answer the question of whether the system in question is adapted or optimised, or if it has been transformed into another system. But what if we improve networked infrastructure by adding a whole class of components, by linking it to new networks, or by cutting linkages? We could think of cutting, for example, the connections between national power grids, in order to enhance resilience (reducing the risk of cascading failures). In the outcome, the technical artefacts may continue to fulfil their function, namely providing energy to the same users as before. However, from a system's point of view, it would not be the same; the system would have been transformed or divided into subsystems.

Bourbeau (2013: 10) has tried to combine the two: resilience should be regarded as an englobing concept which comprises both strategies of continuity and of transformation. Endreß and Rampp (2014: 90-91) propose the concept of 'transformative Autogenese', a kind of third way between the extremes of 'simple continuity' and 'total discontinuity'. For the moment, this may not help us in resolving the problem of technical infrastructure, but it may signpost the right way.

4 Open Questions and Problems

As already described, the benefits of resilience may turn out to be its weaknesses. It is an interdisciplinary concept used across varied disciplines by various practitioners. From a political perspective, resilience is also the subject of power relations and changing forms of governmentality. Like the concept of vulnerability, resilience is a hybrid concept which is descriptive and normative (or even prescriptive) at the same time. As with every normative concept, one should be very careful when using it as an analytical tool. Resilience strategies very often also entail improvements that go beyond sheer technical optimisations but include additional social or cultural values, like more diversity, more participation, social justice or sustainability.

Social and cultural values have an impact on the conception of resilience. Obviously, recent debates tend to favour the evolutionary definition, 'bouncing forward' rather than 'bouncing back'. But the discussion is far from finished. However, using the term resilience runs the risk of obscuring the fact that there are actually at least two opposite concepts inscribed in it, which depend on different visions of how society *should* act or improve.

When using resilience as an analytical concept, the definition of system boundaries seems crucial. Especially when considering the panarchy model by Gunderson and Holling (2002), analysing the resilience of a system on one spatial or organisational scale may lead to neglecting the resilience on a different spatial or organisational scale. Viewing resilience as a process over time also leads to the problem of defining a starting point for the analysis, potential tipping points as well as recovery phases. This might be problematic with an historic view of past systems as well as when analysing current systems and transitions which are not yet terminated.

As mentioned above, the openness of the concept of resilience is an advantage as well as a drawback. Whereas this feature is more or less common to all of the concepts discussed here, there is another aspect which applies especially to resilience, namely the metaphorical character of the verb to 'bounce', which is an integral part of our understanding of resilience. 'Bouncing back' or 'bouncing forward' stimulate the imagination of the reader much more than an abstract definition. It implies elasticity and a kind of smartness, without explicitly naming that. The metaphor probably also helped to make resilience so popular in the public and among scientists. Metaphorical language has the advantage that it appeals to images or experiences that are common to everybody, so that scholars from various disciplines and various intellectual backgrounds may quickly find a common ground of understanding. The disadvantage may be that the metaphor is stronger than a scientific definition, leading to imprecision in the use of the concept.

Empirical social scientists might highlight another problem: even if technical systems possess the capacity to bounce forward or to bounce back in the sense of keeping a defined function (such as providing energy), the community of users may change due to changing economic or social conditions. Consider as an example the debate on 'splintering urbanism'. This debate has pointed to the problem that by privatising urban supply systems, huge parts of the population in Third World metropolises may be excluded from access to basic services. From a systems approach (the city as a social-technical system) one might consider these developments a form of successful adaption to new financial and economic conditions. Alternatively, one might consider it the end of urbanity as embodying the promise of social equality and integration (cf. for splintering urbanism Graham/Marvin 2001). Taking this view, one has to ask the questions of resilience of what and resilience for whom, including questions of social justice (cf. Davoudi et al. 2012). Existing power structures may well lead to a growing divide both in accessibility and in the capacity to resist, recover and adapt.

In conclusion, the multiple definitions and understandings of resilience have to be dealt with. This would mean the concept surpassing the engineering and technocratic domains towards more future-oriented approaches of adaptability, preparedness and complexity management in the light of perceived risks and uncertainty. Nevertheless, it seems that interdisciplinary research combining different understandings of resilience is still in its infancy.

Preparedness & Prevention

Arturo Crespo, Marcus Dombois, Jan Henning

As the notion of resilience with all its adaptive and restoring features gains particular attention, another concept lurks in the back, namely the perspective of a continuously prepared and ready to respond system. In contrast to the other concepts, preparedness and prevention notions are mostly utilised within more mundane discussions. They are explicitly part of solution-oriented practices or schemes in infrastructure engineering and governance. As such, they are not necessarily (but occasionally) connected to debates over conceptual frameworks of these practices. We do think that, starting from mundane pragmatics, it is worth taking them seriously and considering their epistemic implications.

As a general understanding, preparedness & prevention emphasises the action of preventing harmful events from happening and the state of being prepared for their occurrence. With regard to the scientific discourses within social, engineering and natural sciences, one can observe clear tendencies and distinctions in their usage of the concepts.

In the framework of contemporary critical infrastructure research and Emergency Management Studies, preparedness & prevention strategies are to be regarded not only as a forecasting process, conducted prior to the occurrence of a disruptive event, but as a continuous self-adjusting process that enables systems to systematically grow in parallel to their complexity – reflecting once again many features of the debate on resilience. In this sense, preparedness & prevention measures/strategies constitute holistic coping mechanisms against any disruptive event as they support a certain degree of operational continuity, and expand the system's adaptable boundaries, allowing for an exploration of new operational levels.

1   Ad-Hoc Description

Preparedness and prevention have been largely explained as two self-contained notions (Hémond/Robert 2012; UNISDR 2009). One can claim they are contingent on each other. In general,

- preparedness would reduce the impact of harmful events by mitigating its effects, while
- prevention ideally prevents these events from happening in the first place.

Initially, preparedness considers the tacit acceptance and understanding that one cannot prevent everything from occurring (Comfort et al. 2010: 3). As a concept it has been largely described as the set of aptitudes built within a system, to respond and further manage the consequences of imminent and multi-scale disastrous events (Collier/Lakoff 2008b; Comfort et al. 2010; UNISDR 2009). Prevention, on the other hand, is mainly regarded as the foregoing process taken to hinder the development of an adverse situation.

To this day, both concepts have been embedded within the broader emergency management paradigm, and further developed as adept tools, which ultimately constitute the core of disaster response strategies. In the practice of disaster management, they are used conjointly. Therefore, we contend that preparedness & prevention are hard to be considered separately. As King argues: '[P]reparation is entirely contingent upon the scale of the disaster, which is presumably affected by preventive measures that have been put in place. They cannot exist as separate activities.' (King 2007: 658). To underline this, we prefer to elucidate both together as a conjoint concept of 'preparedness & prevention'.

2 Literature Overview

Historian Nicolai Hannig has outlined several key elements of the history of the concept of prevention. The author explained that a precondition for prevention was the perception of nature as a changeable entity. Preventive thinking also implied that a state without catastrophes is the normal state (Hannig 2015: 64). As mentioned above, preparedness & prevention share a glance towards the future. They are an active attempt at changing the future in contrast to passive awaiting. Both concepts gained popularity during the nineteenth century, at the end of which the idea of prevention started to replace the 'preparedness paradigm'. They were both connected to processes like secularisation, rationalisation and nationalisation (Hannig 2015: 34-39). Preparedness & prevention measures helped create the idea and practical reality of states as institutions of emergency response. Such activities were cornerstones of the development of nation states. According to Hannig it is also necessary to be cognisant of the fact that prevention can also be precarious. Often prevention excludes, discriminates and it may well destroy parts of nature. Moreover, it creates power relations, as it sanctions and controls actions

(Hannig 2015: 40-42). Power relations may become obvious in the course of prevention & preparedness measures depending on how they prioritise what is worth protecting.

The concept of preparedness & prevention was further developed in the theatre of military operations throughout the 20th century. Collier and Lakoff mention the emergence of the distributed preparedness policy in the US during the early stages of the Cold War. US planners discussed it in terms of civil defence programmes, which were targeted at preparing the country for a possible nuclear attack (Collier/Lakoff 2008b: 7). Key elements of distributed preparedness plans were emergency federalism and vulnerability mapping.

Later it was introduced into the area of civilian affairs. At this time, the concept of preparedness & prevention became strongly connected to disaster management and other areas within civil security. Disaster and emergency management gained more and more importance during the second half of the twentieth century. In the US, Germany and Canada, the concept of civil security, or public safety as it was called then, was introduced during the 1960s and '70s. The foundation of the Technisches Hilfswerk (THW) during the 1950s in Germany, the building of a legal framework in Canada in the 1960s and the emergence of the Federal Emergency Management Agency (FEMA) in 1979 can be seen as important cornerstones (Hémond/Robert 2012: 405).

At the time, experts underlined the necessity for communities and actors to prevent and prepare better for disasters due to growing dependencies and interdependencies of systems. An increasing number of harmful events identified in the early development of the concept are of particular importance today. In recent years, the discourse about the state of preparedness has shifted in favour of a more flexible stance, the state of resilience. The focus of interest differs between disciplines and organisations. It may well be that some researchers underline response capability while others emphasise the operational continuity (Hémond/Robert 2012: 404).

Public and political debates have heavily influenced the understandings of preparedness & prevention. First, Cold War planners already considered that enemy attacks would focus on industrial and urban centres. In due course they also deemed cities to be major hubs for what are now called critical infrastructures – chief elements worth protecting. Therefore, critical infrastructure protection schemes have always been considered with respect to the strategies of preparedness & prevention.

Second, disaster management was influenced by the most pressing problems at different moments in history, resulting in trends such as more or less centralisation (cf. Comfort et al. 2010). In traditional strategies of preparedness, local knowledge mattered most, resulting in a bottom to top approach. Cold War civil defence schemes accorded a big role to local, state and federal governments. The terror attacks of 9/11 and the state's new focus on prevention of terrorism tends to legitimise the central state's responsibility and influence. The concept of resilience might strengthen a bottom-up approach once again.

Today, preparedness & prevention appear as part of the four phases of emergency management. The Canadian Standards Association calls them prevention, preparedness, intervention and recovery, while another classification would be mitigation, preparedness, response and recovery. Here once again, mitigation puts a strong emphasis on preventing emergencies or reducing their harmful impact while preparedness means preparation to handle an emergency (UNISDR 2009).

Since 2000, the terms 'prevention' and 'preparedness' have appeared in publications of various practical and theoretical fields. They range from crime and terror, to disease prevention and public health preparedness to business continuity and resilience (Hémond/Robert 2012).

As mentioned above, we consider preparedness & prevention as a conjoint concept. However, Collier & Lakoff have made a difference between the two strategies of 'preparedness' and 'precaution' (as they call it). In their eyes, preparedness would be an American concept, and precaution a European one, heavily influenced by the risk theory of Ulrich Beck (Beck 1992). Precaution would be the strategy of avoiding risks by preventing possible developments altogether (for example prohibiting genetically engineered plants). They continue: 'Preparedness does not seek to prevent the occurrence of a disastrous event but rather assumes that the event will happen. Instead of seeking to constrain action in the face of uncertainty, it turns potentially catastrophic threats into vulnerabilities to be mitigated' (Collier/Lakoff 2009: 263).

3 Preparedness & Prevention in Critical Infrastructure Research

We contend that preparedness & prevention should be considered as a conjoint concept, presuming that disruptive events in the running of networked infrastructure are unavoidable. Both take on a very active stance towards the future and consider the uncertain and constantly shifting nature of the world (Hannig 2015), highlighting the ambition of enhancing an overall system's ability to respond to the occurrence of random unexpected events (cf. Collier/Lakoff 2009: 263). In fact, there are at present virtually no important infrastructure systems without both preparedness and prevention qualities. Operators and managers are interested in proactively handling unforeseen hazards and their potential impact on the system's ability to fulfil its objectives. Furthermore, it is very difficult to distinguish a divide between preparedness and prevention, since any prevention strategy influences possible preparative measures. However, in theory one could imagine that there are strategies of prevention that completely lack elements of preparedness, i.e. mitigation strategies in case of harmful impacts. However, from all that we know about the system properties of technical infrastructure, a concentration on sheer prevention would hardly be successful.

With respect to urban critical infrastructure systems, the concept of preparedness & prevention builds upon the socially and technically induced criticality, and acknowledges the need for ways to prevent and respond to imminent failures or breakdowns of these systems. It links vulnerability and resilience to critical infrastructures by bridging the gap of malfunction and contributes to the broader aim of resilience of the system's ever-growing complexity. The concept allows risks and harmful events that impinge on the system from the outside (e.g. natural disasters) as well as those that originate from within (e.g. 'normal accidents') to be coped with (cf. chapter 4). The concept helps organise cooperation and communication between different stakeholders facing an emergency scenario. Furthermore, it also strengthens interdisciplinary perspectives, which is crucial for critical infrastructures such as complex socio-technical systems.

The concept supports the recognition of an increasing number of dependencies and interdependencies between infrastructural systems and realises that there is an increasing need for dynamic strategies in order to account for cascading effects and potential functional system crises. Additionally, preparedness & prevention has a strong link to temporal and spatial relations. While in situations of emergency a quick deployment of response protocols (preparedness) is crucial, prevention also aims at slowing down the time and scope of the distribution of harmful events. Alongside the relationship to time and space, preparedness & prevention is also strongly connected to questions of power and by extension modernity and technical evolution. All of these relations define the characteristics of preparedness & prevention strategies in part.

The construction of a dyke can be a practical example: it is always planned with regard to space and time and functions in its basic sense as a prevention measure against, for example, a 50-year flood event. For a 100-year flood event however, the dyke can also be seen as a preparedness measure and forms part of a wider strategy that also includes, for example, the deployment of mobile walls. Furthermore, since dykes are very expensive structures, site selection is often linked to power politics.

If taken seriously, preparedness & prevention strategies will lead to a dynamic concept, mediating processes of adaptation and transformation, as exemplified in the concept of resilience. So preparedness & prevention strategies are aimed at bridging the system's dynamic vulnerabilities with their ability to foster transforming or adaptive capacities. Preparedness & prevention can help achieve orderly transitions from response to recovery, and allow systems to transform and adapt, thus increasing their resilience level.

4 Conclusion

It is through preparedness & prevention strategies that infrastructure systems are able to cope with harmful events. Recognising that a system is never able to handle all its vulnerabilities, the occurrence of disruptions has to be accepted as unavoidable. Therefore, isolated strategies of mere prevention are not convincing. Within the context of critical infrastructures and by taking time and space into consideration, preparedness & prevention strategies are inextricably linked, as evidenced in the dyke example in section 6.3. However, further research is needed in order to elucidate the pertinence of the strategies' conjoint character systematically.

A second remaining problem is that of the 'costs' of preparedness & prevention. Together with vulnerability assessment, preparedness & prevention enables actors to design measures throughout all phases of a disaster. While each action of prevention and/or preparation can diminish the negative effects of a hazardous event, it still creates drawbacks. However, practicable solutions will have to ensure that one does not get lost in assessing this negative spiral. Implementation of preparedness & prevention requires a high degree of reflexiveness and further research. By the same token it is highly relevant to answer the question which system or which system component is relevant enough to justify complex and/or costly preparedness & prevention measures.

Moreover, preparedness & prevention can have important political implications. History demonstrates that some of these strategies increased state power and helped create the idea of states as civil protection agencies. It remains imperative to keep expanding the reach of the questions that focus on the implication of these measures within the criticality discourse. Nonetheless, preparedness & prevention actions still gain great significance within critical infrastructure debates since they primarily secure the systems' adaptive and transforming capacities, while maintaining minimum or basic operating qualities for the system.

Relations between the Concepts

All Authors

In this chapter we try to elucidate the relations between our concepts, as the respective debates often overlap. We have decided to summarise them all in one chapter. Of course, we could have organised this book differently. We could have described the relations between the concepts in each of the past chapters, taking as our basis a hierarchical view focusing on the perspective of the chapter's concept as 'looking' at the others. This is important, because when starting from the concept of criticality for example, there will be another vision of vulnerability, and the other way round. In order to simplify the structure of our text and to avoid repetitions, we have decided to merge the differing perspectives into one chapter. However, readers will find in some instances juxtapositions of the differing viewing directions.

Before presenting bilateral relationships we want to point out a historical and conceptual commonality. All concepts can be traced back in one or another respect to the context of Cold War (and later on: anti-terrorism) security thoughts. This has been made explicit for criticality (Folkers 2012), vulnerability mapping and resilience (Collier/Lakoff 2008a), preparedness & prevention (Collier/Lakoff 2008b) and, of course, by the overall importance of securitisation thoughts for all concepts. Although the impact of Cold War thought on the epistemology of early disaster research has been elucidated (cf. Stehrenberger 2014), these joint origins have not yet been explored systematically. Future research should try to analyse in which way the social reality of the military (hierarchies, chains of command, demarcation from civil society) might have influenced the concepts. It might transpire that 'classical modern' thought permanently lurks through concepts we consider 'postmodern', such as resilience.[1]

1 Criticality and Vulnerability

Egan (2007: 5) links criticality to the inducement of vulnerability. Because *critical* components are the ones that steer whole systems into a crisis if disrupted,

---

[1] Many thanks to Sabine Höhler for this suggestion.

© Springer Fachmedien Wiesbaden GmbH, part of Springer Nature 2018
J. I. Engels (Ed.), *Key Concepts for Critical Infrastructure Research*,
https://doi.org/10.1007/978-3-658-22920-7_6

they carry certain vulnerabilities. Identifying those infrastructures that are inherently linked to the proper continuation of a society can make vulnerable spots visible. Criticality and vulnerability, if assessed, are features that can hint at the attention that a certain component requires with respect to protection strategies: a function failure due to a vulnerability in a most critical component affects the entire system or society more than a vulnerability in a less critical component. The more a component is critical and vulnerable at the same time, the more it needs attention in protection activities. Lesser attention is needed when a component is most critical but less vulnerable or less critical and most vulnerable. If criticality *and* vulnerability are on a low level, minimal attention is required.

The concepts of vulnerability and criticality are very closely interconnected. This is particularly true when criticality is used in the sense of deficiency or danger. In this case, the criticality of an infrastructure depends on both its relevance for the group of users and the vulnerability of the group of users resulting from this (cf. the Bundesministerium des Innern definition of criticality). Following Alexander Fekete, criticality indicates the relevance of an element or a system for a society; vulnerability determines the impact of an interruption (Fekete 2013). Almost the same applies if we focus on a technical infrastructure's compounds in a function-oriented perspective: the criticality of a compound to a technical system is dependent upon its relevance for the whole system and its vulnerability.

In addition, we can put it the other way round: the relationship between criticality and vulnerability may be determined considering the three factors of vulnerability: exposition, sensitivity so stress and adaptive capacity (cf. ad-hoc description). All of these factors are profoundly affected by the attributed degree of criticality.

Yet another point may be considered. If we adopt the pragmatic concept of criticality, we have to take into account perceptions and ascriptions of criticality and vulnerability as a kind of 'system of communicating tubes'. Sara Bouchon underlines this aspect (Bouchon 2006: 42-43): the social and political construction of criticality is highly dependent upon perceptions of a society's vulnerabilities, and the other way round.

2 Criticality and Resilience

The link between criticality and resilience is not as obvious as the link between the other key concepts is. However, there is a similarity to the relationship between criticality and vulnerability: the more a component or a technical network is critical for a technical or social system, the greater is its impact on the overall system's resilience, be it positive or be it negative. This is an important point for infrastructure resilience research: if we want to assess the resilience of a society or an infrastructure, the most critical parts or systems have to be analysed more closely than the less critical ones, regardless of whether we use a deficiency-oriented or capacity-oriented approach to criticality. For our research topic, this is of utmost importance: the resilient city is inconceivable without resilient urban networked infrastructure. There can be no sensible comprehensive urban development strategy aiming at resilience that would exclude or omit considering technical infrastructure.

This contention may lead to other reflections. First, resilience talk in urban development policy can be used by infrastructure planners and operators to put forward their interests. Resilience is much more linked to positive values and concepts for a better future than criticality is (not necessarily as we have demonstrated in chapter 2). Linking critical infrastructure to resilience in the political process may shed new light on infrastructure. It may enable the stakeholders of technical infrastructure to increasingly link their concerns to future improvements rather than to fears of breakdown and supply shortage after a terrorist attack, for instance. Put simply: in a political context, linking infrastructures to resilience may improve their image and may open up new opportunities for their development.

Second, from the perspective of resilience thought, criticality may help solve the problem of identity. When using the 'bouncing forward' metaphor, an adapting system might transform so thoroughly that it becomes another one. The problem is to determine the criteria for the identity of a system. Of course, there is no simple solution for that problem. But when we focus on the resilience of a city, we might first establish which technical infrastructures are the most critical. We may, subsequently, define the city's identity according to the services of these infrastructures. As long as the services (e.g. public transportation, water sewage, fresh water supply, electricity supply, ...) are provided, even to a limited extent, the city as a system has maintained its identity – if not, the system has changed, for better or for worse.

This consideration may be applicable for the state of emergency, or for wartimes, or for planned transformations, e.g. possible ban of private vehicles in city centres. This may be useful in a historical perspective, too. So we could argue that the city of Rome has maintained its identity from the Antiquity by upholding freshwater provision using antique water pipes without interruption to present times.

However, if we take into consideration social and cultural aspects, the picture becomes more complicated. The restoration of the city of New Orleans after Hurricane Katrina, for instance, may have resulted in the restoration of the services mentioned above in many areas. Instead, many neighbourhoods, typically when socially precarious, did not recover and the former population left the city because they could not afford the additional costs.

3   Criticality and Preparedness & Prevention

There is a close relationship between criticality and the conjoint concept of preparedness & prevention. Preparedness & prevention strategies should focus on the most critical components, or the most critical infrastructures, first. As we define a high level of criticality as a high level of interrelatedness, protection and mitigation strategies will be more effective when they start by considering the most critical elements. This is nothing more than a pragmatic procedure, reducing the complexity of the task of protection by identifying the 'biggest construction sites'. Criticality can be used to prioritise the individual components and distribute resources towards preventing a disruption or being prepared for it.

Here we can go back to the paper of Sara Bouchon, who recommends calling critical those systems and services which help to cope with a crisis or attenuate the effects of a critical event (Bouchon 2006: 37). This usage of the term creates a certain ambivalence with respect to vulnerability, because it implies the vulnerable as well as the non-vulnerable infrastructural systems. At the same time, this usage reflects the growing importance of the concept of preparedness, because it implies that at least parts of the systems are prepared for (or are preparing society for) the moment of malfunction of other parts. Bouchon's proposition only makes sense in the context of those efforts which aim at accepting possible infrastructure breakdown by providing resilient auxiliary systems.

However, the liaison between preparedness & prevention and criticality is dependent on the context. This is particularly true for systems which are constantly targeted by criticality labelling actions. The crux of the matter resides in deciding, from a function-oriented criticality standpoint, which of these criticality attributions are convincing.

On the other hand, the implementation of any preparedness & prevention plan itself contributes to the process of attributing criticality to the components or systems which will be objects of preparedness & prevention. The aim and focus of preparedness & prevention measures necessarily derives from a general categorisation, which gives priority to some infrastructural systems, their hazard-related management and their coping mechanisms over others. This consideration makes clear that criticality, either as a function-oriented or a pragmatic concept, is always an outcome of specific choices made by experts or decision makers. Ascribing criticality to an infrastructure is dependent on a broader framework of values and power relations (Metzger 2004b). From a theoretical point of view, criticality ascriptions on the one hand and preparedness & prevention schemes on the other mutually confirm themselves or at least are dependent on one another.

In the urban context, complex infrastructural systems are of utmost importance and cities tend to make important efforts to prepare for and prevent breakdowns. The simple reason for this is that they are considered as prerequisites of urban lifestyles and at the same time remain among the chief hazard and vulnerability sources within cities. As a consequence, it has been relentlessly argued that plans and protocols aimed at reducing, addressing or controlling these self-generated threats are essential (Godschalk 2003). Therefore, preparedness & prevention strategies must be inherently linked to the forces behind the categorising implications, in which a system can be considered to remain more critical or relevant than another. As such, the impact of a system's breakdown is not only perceived by its actual implications, but the population's subjective perception built around the existing discourses becomes relevant as well. Hence, the extent to which the impact of a certain hazard on any of the systems' elements is quantifiable ceases to be the leading factor of importance (Metzger 2004b). Instead, preparedness & preventions actions can be taken with different objectives and adapted to fit any categorisation or scenario. Ultimately, just like criticality, preparedness & prevention resides at the outset of power structures, where the way potential crises are handled can be capitalised in a broad range of ways.

4 Vulnerability and Resilience

Two different research communities, each concerned primarily with vulnerability *or with* resilience, have been working more or less separately from each other. However, in contrast to other relations between concepts discussed in this text, the liaison between resilience and vulnerability has often been addressed in existing literature by authors who work with both concepts. They call for collaboration and exchange between these communities. Many arguments have been put forward to show the potential of combining the research traditions of two concepts (Turner 2010: 574; Christmann/Ibert 2012: 26; Gallopín 2006). Fiona Miller et al. (2010) have systematically examined the main differences and commonalities of the two concepts and the respective scientific communities, so that one can start easily from this work.

While both concepts are mainly concerned with a given system's reaction to stress (Miller et al. 2010: n.p.), they entail different perspectives. However, there is no common opinion about their exact relationship. While Carl Folke et al. consider them as a pair of opposites (2002: 13), others describe resilience rather as resulting from vulnerability or even as to be considered when vulnerability is assessed (cf. Miller et al. 2010: n.p.; Gallopín 2006: 301).

Considering both as counterparts seems quite intuitive. An increase in resilience should be thought of as associated with a reduction in vulnerability, and vice versa. But if vulnerability is defined as the product of sensitivity, exposure, and coping or adaptive capacities, assuming it to be the flipside of resilience falls short of accounting for exposure and sensitivity (Turner et al. 2003). In another part of the literature, resilience is instead integrated into the concept of vulnerability. Here resilience is used to determine or to describe the degree of the coping or adaptive capacities of a system (which are part of the vulnerability concept). In this case, the relationship between adaptive capacities and resilience has to be explained very carefully (cf. Gallopín 2006: 301).

5 Vulnerability and Preparedness & Prevention

From the perspective of vulnerability research, it can be stated that both aspects of preparedness & prevention are concepts which describe strategies to deal with vulnerability.

Understanding the (causes of the) vulnerability of a specific technical infrastructure may be considered as a prerequisite for developing potential measures in the field of preparedness or prevention. It may not be useful trying to prevent or prepare for a system's failure if one does not know to what extent it is vulnerable, which kind of risks seem the most serious, etc. This assertion can be substantiated by critical infrastructure protection history. During the Cold War's first years, the newly developed technique of 'vulnerability mapping' pinpointed the most critical and vulnerable spots which needed to be protected against potential enemy attacks (Collier/Lakoff 2008b: 12). However, the importance of vulnerability assessments has changed since that time. It depends on a large number of different qualities (exposure, sensitivity, adaptation) of the system in question. When we change perspective, however, we get a different image: if an infrastructure's flexibility or learning capacities are highly developed, the need for clearly identifying its vulnerabilities may be substantially reduced (cf. positive conception of vulnerability).

Presupposing that vulnerability is defined by exposure, sensitivity and adaptive capacities, it can be said that preparedness and prevention have a slightly different focus on the different factors of vulnerability. While prevention would be an answer to (high) exposure, strategies of preparedness focus on the enhancement of adaptive capacities. The reduction of sensitivity is addressed by both (i.e. preparedness and prevention) to the same extent. Taken together, preparedness and prevention are dealing with vulnerabilities in a holistic way.

From the standpoint of preparedness & prevention measures, vulnerability (conjointly with criticality) has been established as a tool to identify or pinpoint specific components within a system that need preparedness & prevention mechanisms. As described by Adger, vulnerability has been purposely mediated so as to bring to light states of susceptibility, and ultimately channel the necessary actions to reduce risk (Adger 2006: 268).

Moreover, it transpires that vulnerability (defined as including the adaptive capacities of a system) facilitates the understanding of a system's abilities and reactions to stress. It paves the way to conceptualising how a system interacts with stress or disasters. Describing these (possible) interactions is an important precondition for developing strategies of preparedness – we could even argue that without vulnerability analysis no sensible preparedness & prevention strategy is conceivable.

However, preparedness & prevention strategies are today no longer mere sets of rigid protocols prescribing how an infrastructural system may respond to hazards. Rather, transforming and adaptive capacities are put forward. Hence, and as supported by Adger's research, preparedness & prevention strategies can no longer be characterised as being purely responsive or being inversely proportional to vulnerability. They constitute the foundations of a dynamic process (Adger 2006).

Moreover, as has been detailed in chapter 6, preparedness & prevention measures may increase, as a side effect, certain vulnerabilities in a system while reducing others. As a consequence, we consider it necessary never to lose sight of vulnerability when developing preparedness & prevention strategies. The combination of the two concepts helps in dealing with trade-offs. Vulnerability analysis should be at the beginning and at the end of every preparedness & prevention scheme.

6 Resilience and Preparedness & Prevention

One could argue that resilience, one of the trending terms in contemporary emergency management, is just a more sophisticated approach to preparedness & prevention: in the end it is also about preparing for the disaster and preventing it from happening. However, the most prominent resilience authors focus on adaptive capacities: 'First, the transboundary nature of modern threats widens the range of the contingencies that can besiege a society. Second, modern societies have become increasingly vulnerable to threats new and old. Third, the changing political climate has made it harder for public leaders to deal with crises' (Comfort et al. 2010: 5). Assuming that risks, perturbations and breakdowns have become unavoidable, it appears as if resilience is closer to preparedness than prevention.

Literature has elucidated this problem, with differing results. On the one hand, preparedness & prevention has been portrayed as a set of fixed strategies that need to be guided towards developing a more resilience-related agenda. Here, preparedness & prevention strategies are interpreted as a set of rigid protocols that follow a strict timeline and, as a consequence, they perpetuate the pre-existing character of a system. Thus, it is maintained that due to their static structure they are not able to foster authentic adaptive capacities (Hémond/Robert 2012: 412). This interpretation is conducive towards considering preparedness & prevention strategies as stiff protocols and measures in urgent need of evolving towards a more resilience-related framework – in other words, taking the concept of resilience seriously would help modernise preparedness & prevention.

On the other hand, another group of scholars conceive preparedness & prevention strategies *within* the resilience framework. In their eyes, preparedness & prevention strategies are options which need to be balanced or combined with other options (e.g. response, recovery and even adaptation) (Comfort et al. 2010; Jenelius/Mattsson 2015: 20). In this regard, preparedness & prevention strategies are cornerstones *within* a framework aimed at producing resilience – for example by protecting the integrity of the system's basic capacities enabling adaptation capabilities following shocks (Haimes et al. 2008). Here, preparedness & prevention strategies are conceived as passively and actively adapting the system's ability to cope with hazards, moderating the transition from the 'normal' state of function to a reduced function after the shock. The latter would be described as an 'acceptable level of functioning' (Hémond/Robert 2012: 413). Above all, an 'acceptable level of functioning' lays the groundwork for contemplating failure as unavoidable and stressing the need for adopting clear measures to deal with degraded situations. In the context of critical infrastructures dealing with degraded conditions, this means highlighting the ability to dynamically adapt complex systems to a broad range of unknown states of affairs.

Preparedness, defined as a system's capability to adapt and to transform after a shock, may be more than just securing the lowest acceptable level of functioning. If conceived well and in specific cases, it may lead to a new and better functioning of a system (this is the other side of the coin: not only new vulnerabilities, but also new capacities might arise from these strategies). Under the condition that we conceive resilience as the capacity to 'bounce forward', preparedness may essentially contribute to resilience as it may allow for a fundamental transformation. If so, the abovementioned problem of a system's identity returns: where are the limits of a system's transformation, at which level would we have to speak of a new system?

Bearing that in mind, the concept of resilience including the identity problem, together with the concept of (newly induced) vulnerability, may provide a framework for preparedness & prevention planners to reflect upon the consequences of their strategies.

Bibliography

Adger, W. Neil (2006): Vulnerability. In: Global Environmental Change, 16(3), 268–281.

https://doi.org/10.1016/j.gloenvcha.2006.02.006

Alloy, Lauren B./Abramson, Lyn Y. (2005): Cognitive Vulnerability to Depression. In: Freeman, Arthur et al. (Ed.): Encyclopedia of Cognitive Behavior Therapy. London: Springer, 126–129.

https://doi.org/10.1192/bjp.176.3.302-a

Aradau, Claudia (2010): Security That Matters: Critical Infrastructure and Objects of Protection. In: Security Dialogue 41, 491–514.

http://doi.org/10.1177/0967010610382687

Aradau, Claudia (2014): The promise of security: resilience, surprise and epistemic politics. In: Resilience. International Policies, Practices and Discourses 2, 73–87.

http://doi.org/10.1080/21693293.2014.914765

Bach, Claudia/Bouchon, Sara/Fekete, Alexander/ Birkmann, Jörn/Serre, Damien (2013): Adding value to critical infrastructure research and disaster risk management: the resilience concept. S.A.P.I.EN.S. Surveys and Perspectives Integrating Environment and Society Online

Bak, Per/Paczuski, Maya (1995): Complexity, Contingency, and Criticality. In: Proceedings of the National Academy of Sciences 92, 6689–6696.

https://doi.org/10.1073/pnas.92.15.6689

Bankoff, Greg/Frerks, Georg/Hilhorst, Dorothea (2004): Mapping Vulnerability. Disasters, Development and People. London: Earthscan Publications.

https://doi.org/10.4324/9781849771924

Balzacq, T. (2005): The Three Faces of Securitization: Political Agency, Audience and Context. In: European Journal of International Relations, 11(2), 171–201.

Beck, Ulrich (1992): Risk society: Towards a new modernity. Theory, culture and society. London: Sage.

Berking, Helmuth/ Löw, Martina (2008): Die Eigenlogik der Städte. Neue Wege für die Stadtforschung. Frankfurt a. M., New York: Campus.

https://doi.org/10.1007/978-3-531-94112-7_14

Beyer, Jürgen (2006): Pfadabhängigkeit. Über institutionelle Kontinuität, anfällige Stabilität und fundamentalen Wandel. Frankfurt a. M.: Campus.

Bijker, Wiebe E./Hughes, Thomas P./Finch, Trevor J. (Eds.) (1987): The Social Construction of Technological Systems. Cambridge: MIT Press.

Bijker, Wiebe E./Hommels, Anique/Mesman, Jessica (2014): Studying Vulnerability in Technological Cultures. In: Bijker, Wiebe E./Hommels, Anique/Mesman, Jessica (Eds.): Studying Vulnerability in Technological Cultures. New Directions on Research and Government. Cambridge, MA: MIT Press, 1–26.

https://doi.org/10.1093/scipol/scv001

© Springer Fachmedien Wiesbaden GmbH, part of Springer Nature 2018
J. I. Engels (Ed.), *Key Concepts for Critical Infrastructure Research*,
https://doi.org/10.1007/978-3-658-22920-7

Birkmann, Jörn (2011): Indikatoren zur Abschätzung von Vulnerabilität und Bewältigungspotenzialen. Am Beispiel von wasserbezogenen Naturgefahren in urbanen Räumen. Bonn: BBK-Verlag.

Birkmann, Jörn (2013): Measuring Vulnerability to Promote Disaster-Resilient Societies. Conceptual Frameworks and Definitions. In: Birkmann, Jörn (Ed.): Measuring Vulnerability to Natural Hazards. Towards Disaster Resilient Societies. Tokyo: United Nations Univ. Press, 9–53.

Bouchon, Sara (2006): The Vulnerability of Interdependent Critical Infrastructures Systems. Epistemological and Conceptual State-of-the-Art (No. EUR 22205 EN). Luxembourg: Institute for the Protection and Security of the Citizen.

Bourbeau, Philippe (2013): Resiliencism. Premises and Promises in Securitisation Research. In: Resilience. International Policies, Practices and Discourses 1, 3–17.

Brassett, J. / Vaughan-Williams, N. (2015): Security and the performative politics of resilience: Critical infrastructure protection and humanitarian emergency preparedness. In: Security Dialogue, 46(1), 32–50.

Brklacich, Mike/Bohle, Hans-Georg (2006): Assessing Human Vulnerability to Global Climatic Change. In: Ehlers, Eckart/Kraft, Thomas (Ed.): Earth System Science in Anthropocene. Emerging Issues and Problems. Berlin, Heidelberg: Springer, 51–61.

https://doi.org/10.1007/b137853

Bürkner, Hans-Joachim (2010): Vulnerabilität und Resilienz. Forschungsstand und sozialwissenschaftliche Untersuchungsperspektiven. Erkner: Leibniz-Institut für Regionalentwicklung und Strukturplanung.

Bundesamt für Bevölkerungsschutz und Katastrophenhilfe (2017): Deutschsprachiges Glossar - Kritikalität. Retrieved November 14, 2017.

https://www.kritis.bund.de/SubSites/Kritis/DE/Servicefunktionen/Glossar/Functions/glossar.html?lv2=4968594

Bundesministerium des Innern (2009): Nationale Strategie zum Schutz Kritischer Infrastrukturen. Berlin: Bundesministerium des Innern.

Buzan, Barry/Waever, Ole/Wilde, Jaap de (1998): Security. A New Framework for Analysis. Boulder: Rienner.

Chambers, Robert (1989): Editorial Introduction: Vulnerability, Coping and Policy. In: IDS Bulletin, 20 (2), 1–7.

Chandler, David (2014): Beyond neoliberalism: Resilience, the new art of governing complexity. In: Resilience, 2 (1), 47–63.

http://doi.org/10.1080/21693293.2013.878544

Chandler, David/Coaffee, Jon (Eds.) (2017): Routledge handbook of international resilience. London: Routledge.

Chelleri, Lorenzo (2012): Lessons for Urban Resilience from Delta Urbanism: The Dutch Polders Case. In: Chelleri, Lorenzo/Olazabal, Marta (Eds.): Multidisciplinary perspectives on urban resilience. A workshop report. Bilbao: BC3, 45–51.

Chelleri, Lorenzo/Waters, James J./Olazabal, Marta/Minucci, Guido (2015): Resilience Trade-Offs. Addressing Multiple Scales and Temporal Aspects of Urban Resilience. In: Environment & Urbanization, 27 (1), 181–198.

http://doi.org/10.1177/0956247814550780

Christmann, Gabriela B./Ibert, Oliver (2012): Vulnerability and Resilience in a Socio-Spatial Perspective. A Social-Scientific Approach. In: Raumforschung und Raumordnung, 70 (4), 259–272.

https://doi.org/10.1007/s13147-012-0171-1

Christmann, Gabriela B./Ibert, Oliver/Kilper, Heiderose/Moss, Timothy (2011): Vulnerabilität und Resilienz in sozio-räumlicher Perspektive – Begriffliche Klärungen und theoretischer Rahmen, Erkner: IRS Working Paper (44).

Coaffee, Jon (2009): Terrorism, risk and the global city: Towards urban resilience. Farnham: Ashgate.

https://doi.org/10.1093/pa/gsr050

Coaffee, Jon/Clarke, Jonathan (2015): On securing the generational challenge of urban resilience. In: Town Planning Review, 86 (3), 249–255.

Coaffee, Jon/Clarke, Jonathan (2016): Critical infrastructure lifelines and the politics of anthropocentric resilience. In: Resilience, 1–21.

http://doi.org/10.1080/21693293.2016.1241475

Collet, Dominik (2012): 'Vulnerabilität' als Brückenkonzept der Hungerforschung. In: Collet, Dominik/Lasse, Thoren/Schanbacher, Ansgar (Ed.): Handeln in Hungerkrisen. Neue Perspektiven auf soziale und klimatische Vulnerabilität. Göttingen: Universitätsverlag, 13–25.

https://doi.org/10.17875/gup2012-476

Collier, Stephen/Lakoff, Andrew (2008a): The Vulnerability of Vital Systems: How 'critical infrastructure' became a security problem. In: Dunn, Myriam A./Kristensen, Kristian S. (Eds.): The Politics of Securing 'the Homeland'. Critical Infrastructure, Risk and (In)Security, London: Routledge 2008, 17–39.

Collier, Stephen J./Lakoff, Andrew (2008b): Distributed Preparedness: The Spatial Logic of Domestic Security in the United States. In: Environment and Planning D: Society and Space, 26 (1), 7–28. https://doi.org/10.1068/d446t

Collier, Stephen J./Lakoff, Andrew (2009): Infrastructure and Event: The Political Technology of Preparedness. In: Braun, Bruce/Whatmore, Sarah (Eds.): The Stuff of Politics. Technoscience, Democracy, and Public Life. Minneapolis: University of Minnesota Press, 243–266.

Comfort, L. K./Boin, A./Demchak, C. C. (Eds.) (2010): Designing Resilience: Preparing for Extreme Events. Pittsburgh, Pa: University of Pittsburgh Press.

https://doi.org/10.2307/j.ctt5hjq0c

Davoudi, Simin/Shaw, Keith/Haider, L. Jamila/Quinlan, Allyson E./Peterson, Garry D. /Wilkinson, Cathy/Fünfgeld, Hartmut/McEvoy, Darryn/Porter, Libby (2012): Resilience: A Bridging Concept or a Dead End? 'Reframing' Resilience: Challenges for Planning Theory and Practice Interacting Traps: Resilience Assessment of a Pasture Management System in Northern Afghanistan Urban Resilience: What Does it Mean in Planning Practice? Resilience as a Useful Concept for Climate Change Adaptation? The Politics of Resilience for Planning: A Cautionary Note [sic!]. In: Planning Theory & Practice, 13 (2), 299–333.

Davoudi, Simin/Bohland, James/Knox, Paul L./Lawrence, Jennifer L. (2017): The Resilience Machine. In: Urban resilience research Net.

Desouza, Kevin C./Flanery, Trevor H. (2013): Designing, planning, and managing resilient cities: A conceptual framework. In: Cities, 35, 89–99.

Dolata, Ulrich/Werle, Raymund (2007): 'Bringing Technology back in': Technik als Einflussfaktor sozioökonomischen und institutionellen Wandels. In: Dolata, Ulrich/Werle, Raymund (Eds.): Gesellschaft und die Macht der Technik. Frankfurt a. M., New York: Campus, 15–43.

Duijnhoven, Hanneke/Neef Martijn (2016): Disentangling Wicked Problems: A Reflexive Approach Towards Resilience Governance. In: Masys, Anthony J. (Ed.): Applications of Systems Thinking and Soft Operations Research in Managing Complexity. From Problem Framing to Problem Solving. Heidelberg: Springer, 91–106.

Egan, M. J. (2007): Anticipating Future Vulnerability: Defining Characteristics of Increasingly Critical Infrastructure-like Systems. In: Journal of Contingencies and Crisis Management, 15 (1), 4–17.

https://doi.org/10.1111/j.1468-5973.2007.00500.x

Endreß, Martin (2015): The Social Constructedness of Resilience. In: Social Sciences, 4, 533–545.

Endreß, Martin/Rampp, Benjamin (2014): Resilienz als Prozess transformativer Autogenese. Schritte zu einer soziologischen Theorie. In: BEHEMOTH - A Journal on Civilisation, 7 (2), 73–102.

Engels, Jens Ivo (2018): Relevante Beziehungen. Vom Nutzen des Kritikalitätskonzepts für Geisteswissenschaftler. In: Engels, Jens Ivo/Nordmann, Alfred (Eds.) (2018): Was heißt Kritikalität? Zu einem Schlüsselbegriff der Debatte um Kritische Infrastrukturen. Bielefeld: transcript.

Engels, Jens Ivo/Schenk, Gerrit J. (2014): Infrastrukturen der Macht - Macht der Infrastrukturen. Überlegungen zu einem Forschungsfeld. In: Förster, Birte/Bauch, Martin (Eds.): Wasserinfrastrukturen und Macht von der Antike bis zur Gegenwart. München: Oldenbourg, 22–58.

Engels, Jens Ivo/Nordmann, Alfred (Eds.) (2018): Was heißt Kritikalität? Zu einem Schlüsselbegriff der Debatte um Kritische Infrastrukturen. Bielefeld: transcript.

Engler, Steve (2012): Developing a Historically Based 'Famine Vulnerability Analysis Model' (FVAM). An Interdisciplinary Approach. In: Erdkunde, 66 (2), 157–172.

European Commission Publications Office (2006): EPCIP - The European Programme for Critical Infrastructure Protection (No. MEMO/06/477). Brussels: European Commission.

Fekete, Alexander (2011): Common Criteria for the Assessment of Critical Infrastructures. In: International Journal of Disaster Risk Science, 2, 15–24.

Fekete, Alexander (2013): Schlüsselbegriffe im Bevölkerungsschutz zur Untersuchung der Bedeutsamkeit von Infrastrukturen – von Gefährdung und Kritikalität zu Resilienz und persönlichen Infrastrukturen. In: Unger, Christoph (Ed.): Krisenmanagement – Notfallplanung – Bevölkerungsschutz. Festschrift anlässlich 60 Jahre Ausbildung im Bevölkerungsschutz, dargebracht von Partnern, Freunden und Mitarbeitern des Bundesamtes für Bevölkerungsschutz und Katastrophenhilfe. Berlin: Duncker & Humblot, 327–340.

Fekete, Alexander/Tzavella, Katerina/Baumhauer, Roland (2016): Spatial Exposure Aspects Contributing to Vulnerability and Resilience Assessments of Urban Critical Infrastructure in a Flood and Blackout Context. In: Natural Hazards, 84 (3), 1–26.

FIRST.org Inc. (2015): Common Vulnerability Scoring System v3.0: Specification Document.

Folke, Carl/Carpenter, Steve/Elmqvist, Thomas/Gunderson, Lance/ Holling, C. S./Walker, Brian (2002): Resilience and Sustainable Development. Building Adaptive Capacity in a World of Transformations. In: Ambio, 31 (5), 437–440.

Folkers, Andreas (2012): Kritische Infrastruktur. In: Marquardt, Nadine/ Schreiber, Verena (Eds.): Ortsregister. Ein Glossar zu Räumen der Gegenwart. Bielefeld: transcript, 154–159.

Gallopín, Gilberto C. (2006): Linkages Between Vulnerability, Resilience and Adaptive Capacity. In: Global Environmental Change, 16 (3), 293–303.

George Mason University, School of Law (2007): Critical Thinking: Moving from Infrastructure Protection to Infrastructure Resilience (CIP Program Discussion Paper Series). Arlington, VA.

Godschalk, D. R. (2003): Urban Hazard Mitigation: Creating Resilient Cities. In: Natural Hazards Review, 4 (3), 136–143.

https://doi.org/10.1061/(ASCE)1527-6988(2003)4:3(136)

Goldstein, Bruce Evan (Ed.) (2012) Collaborative resilience: Moving through crisis to opportunity. Cambridge, Massachusetts: MIT Press.

https://doi.org/10.1080/15575330.2013.811877

Graham, Stephen (Ed.) (2010): Disrupted Cities: When Infrastructure Fails. London, New York: Routledge.

https://doi.org/10.4324/9780203894484

Graham, Stephen/Marvin, Simon (2001): Splintering Urbanism. Networked Infrastructures, Technological Mobilities and the Urban Condition. London, New York: Routledge.

Graham, Leigh/Debucquoy, Wim/Anguelovski, Isabelle (2016): The Influence of Urban Development Dynamics on Community Resilience Practice in New York City after Superstorm Sandy: Experiences from the Lower East Side and the Rockaways. In: Global Environmental Change, 40, 202–212.

https://doi.org/10.1016/j.gloenvcha.2016.07.001

Great Britain Cabinet Office (2010): A Strong Britain in an Age of Uncertainty: The National Security Strategy. London: Stationery Office.

Gunderson, Lance H./Holling, Crawford S. (Eds.) (2002): Panarchy. Understanding Transformations in Human and Natural Systems. Washington: Island Press.

https://doi.org/10.1016/S0006-3207(03)00041-7

Haimes, Y. Y./Crowther, K./Horowitz, B. M. (2008): Homeland security preparedness: Balancing protection with resilience in emergent systems. In: Systems Engineering, 11 (4), 287–308.

https://doi.org/10.1002/sys.20101

Hannig, Nikolai (2015): Die Suche nach Prävention. Naturgefahren im 19. und 20. Jahrhundert. In: Historische Zeitschrift, 300 (1), 33–65. https://doi.org/10.1515/hzhz-2015-0002

Hémond, Y./Robert, B. (2012): Preparedness: the state of the art and future prospects. In: Disaster Prevention and Management: An International Journal, 21 (4), 404–417.

https://doi.org/10.1108/09653561211256125

Hempel, Leon/Lorenz, Daniel F. (2014): Resilience as an Element of a Sociology of Expression. In: Behemoth 7 (2), 26–72.

https://doi.org/10.6094/behemoth.2014.7.2.833

Hodson, Mike/Marvin, Simon (2010): World cities and climate change: Producing urban ecological security. Issues in society. Berkshire: McGraw-Hill.

Högselius, Per et al. (2013): The Making of Europe's Critical Infrastructure: Common Connections and Shared Vulnerabilities. Houndmills, Basingstoke: Palgrave Macmillan.

https://doi.org/10.1111/jcms.12269_9

Höhler, Sabine (2014): Resilienz: Mensch – Umwelt – System. Eine Geschichte der Stressbewältigung von der Erholung zur Selbstoptimierung. In: Zeithistorische Forschungen 11, 425–443.

Holling, Crawford S. (1973): Resilience and Stability of Ecological Systems. In: Annual Review of Ecology and Systematics, 4 (1), 1–23.

https://doi.org/10.1146/annurev.es.04.110173.000245

Hughes, Thomas P. (1987): The Evolution of Large Technological Systems. In: Bijker, Wiebe E./Hughes, Thomas P./Pinch, Trevor J. (Eds.): The Social Construction of Technological Systems. Cambridge, 51–82.

Hughes, Thomas P./Mayntz, Renate (Eds.) (1988): The development of large technical systems. Frankfurt a. M.: Campus.

Jabareen, Yosef (2015): The Risk City (Vol. 29). Dordrecht: Springer Netherlands.

Jasanoff, Sheila (Ed.) (2004): States of Knowledge. The Co-production of Science and Social Order. London, New York: Routledge.

Jenelius, E./Mattsson, L.-G. (2015): Road network vulnerability analysis: Conceptualization, implementation and application. In: Computers, Environment and Urban Systems, 49, 136–147.

https://doi.org/10.1016/j.compenvurbsys.2014.02.003

Johnson, C./Blackburn, S. (2014): Advocacy for urban resilience: UNISDR's Making Cities Resilient Campaign. In: Environment and Urbanization 26 (1), 29–52.

https://doi.org/10.1177/0956247813518684

Joseph, Jonathan (2013): Resilience as embedded neoliberalism: A governmentality approach. In: Resilience 1 (1), 38–52.

http://doi.org/10.1080/21693293.2013.765741

Katina, P. F./Hester, P. T. (2013): Systemic determination of infrastructure criticality. In: International Journal of Critical Infrastructures 9 (3), 211–225.

https://doi.org/10.1504/IJCIS.2013.054980.

Kaufmann, Stefan (2012): Resilienz als ‚Boundary Object'. In: Daase, Christopher/Offermann, Philipp/Rauer, Valentin (Eds.): Sicherheitskultur. Soziale und politische Praktiken der Gefahrenabwehr. Frankfurt am Main: Campus, 109–131.

King, D. (2007): Organisations in Disaster. In: Natural Hazards, 40 (3), 657–665.

https://doi.org/10.1007/s11069-006-9016-y

Klein, Richard J. T./Nicholls, Robert J./Thomalla, Frank (2003): Resilience to natural hazards: How useful is this concept? In: Environmental Hazards, 5 (1), 35–45.

https://doi.org/10.1016/j.hazards.2004.02.001

Koselleck et al. (1982): Krise und Kritik. In: Geschichtliche Grundbegriffe. Historisches Lexikon zur politisch-sozialen Sprache in Deutschland (Vol. 3). Stuttgart: Klett-Cotta.

Kröger, Wolfang/Zio, Enrico (2011): Vulnerable Systems. London: Springer.

Laak, Dirk van (2004): Technological Infrastructure, Concepts and Consequences. In: ICON. Journal of the International Committee for the History of Technology, 10, 53–64.

Lakoff, Andrew/Collier, Stephen J. (2010): Infrastructure and Event: The Political Technology of Preparedness. In: Braun, Bruce/Whatmore, Sarah (Eds.): Political matter. Technoscience, democracy, and public life. Minneapolis: University of Minnesota Press, 243–266.

Latour, Bruno (1993): We have never been modern. Cambridge: Harvard University Press.

Latour, Bruno (2004): Why has Critique Run Out of Steam? From Matters of Fact to Matters of Concern. In: Critical Inquiry, 30 (2), 225–248.

http://doi.org/10.1086/421123

Lenz, Susanne (2009): Vulnerabilität Kritischer Infrastrukturen. Bonn: Bundesamt für Bevölkerungsschutz und Katastrophenhilfe.

Little, Richard G. (2009): Controlling Cascading Failure: Understanding the Vulnerabilities of Interconnected Infrastructures. In: Journal of Urban Technology, 9, 109–123.

https://doi.org/10.1080/106307302317379855

Matyas, David/Pelling, Mark (2015): Positioning resilience for 2015: the role of resistance, incremental adjustment and transformation in disaster risk management policy. In: Disasters, 39 Suppl. 1, 1–18.

McPhearson, Timon/Andersson, Erik/Elmqvist, Thomas/Frantzeskaki, Niki (2015): Resilience of and through urban ecosystem services. In: Ecosystem Services, 12, 152–156.

https://doi.org/10.1016/j.ecoser.2014.07.012

Medd, Will/Marvin, Simon (2005): From the Politics of Urgency to the Governance of Preparedness: A Research Agenda on Urban Vulnerability. In: Journal of Contingencies and Crisis Management, 13 (2), 44–49.

https://doi.org/10.1111/j.1468-5973.2005.00455.x

Metzger, Jan (2004a): Das Konzept 'Schutz kritischer Infrastrukturen' hinterfragt. In: Bulletin zur schweizerischen Sicherheitspolitik, 73–85.

Metzger, Jan (2004b): The Concept of Critical Infrastructure Protection. In: Bailes, A. J. K./Frommelt, I. (Eds.): Business and Security. Public-Private Sector Relationships in a New Security Environment. Oxford: University Press.

Miller, Fiona/Osbahr, Henny/Boyd, Emily/Thomalla, Frank/Bharawani, Sukaina/Ziervogel, Gina/Walker, Brian/Birkmann, Jörn/van der Leeuw, Sander/Rockström, Johan/Hinkel, Jochen/Downing, Tom/Folke, Carl/Nelson, Donald (2010): Resilience and vulnerability: complementary or conflicting concepts? In: Ecology and Society, 15 (3), 1–25.

Mollinga, Peter P. (2008): The rational organisation of dissent: Boundary concepts, boundary objects and boundary settings in the interdisciplinary study of natural resources management. ZEF Working Paper Series, 33. Bonn: ZEF.

Monstadt, Jochen (2009): Conceptualizing the political ecology of urban infrastructures: Insights from technology and urban studies. In: Environment and Planning, A 41 (8), 1924–1942.

https://doi.org/10.1068/a4145

Moteff, J./Copeland, C./Fischer, J. (2003): Critical Infrastructures: What Makes an Infrastructure Critical?. Washington: The Library of Congress.

Murmann, Johann Peter (2003): Knowledge and Competitive Advantage. The Coevolution of Firms, Technology, and National Institutions. Cambridge: University Press.

https://doi.org/10.1057/palgrave.jibs.8400113

Olazabal, Marta/Chelleri, Lorenzo/Waters, James J. (2012): Why Urban Resilience? In: Chelleri, Lorenzo/Olazabal, Marta (Eds.): Multidisciplinary perspectives on urban resilience. A workshop report. Bilbao: BC3, 7–18.

Perrow, Charles (1984): Normal Accidents. Living with High Risk Technologies. Princeton: Princeton University Press.

Pescaroli, Gianluca/Alexander, David (2016): Critical Infrastructure, Panarchies and the Vulnerability Paths of Cascading Disasters. In: Natural Hazards, 82 (1), 175–192.

https://doi.org/10.1007/s11069-016-2186-3

Pizzo, Barbara (2015): Problematizing resilience: Implications for planning theory and practice. In: Cities, 43, 133–140.

https://doi.org/10.1016/j.cities.2014.11.015

Prior, Tim (2015): Measuring Critical Infrastructure Resilience: Possible Indicators (Risk and Resilience Report No. 9). Zürich.

Rinaldi, Steven M./Peerenboom, James P./Kelly, Terrence K. (2001): Identifying, Understanding and Analyzing Critical Infrastructure Interdependencies. In: IEEE Control Systems Magazine, 21 (6), 11–25.

https://doi.org/10.1109/37.969131

Riskind, John H./Black, David (2005): Cognitive Vulnerability. In: Freeman, Arthur et al. (Eds.): Encyclopedia of Cognitive Behavior Therapy. London: Springer, 122–126.

Rodin, Judith (2014): The resilience dividend: Being strong in a world where things go wrong. New York: PublicAffairs.

Ropohl, Günter (2009): Allgemeine Technologie. Eine Systemtheorie der Technik. Karlsruhe: Universitätsverlag, 3rd edition.

Schatzki, Theodore R. (2003): Nature and Technology in History. In: History and Theory, 42, 82–93.

Schenk, Gerrit Jasper (2015): Städte zwischen Vulnerabilität und Resilienz. Kritische Infrastruktur und der gesellschaftliche Umgang mit Katastrophen in der Geschichte. In: Symposium 2015. Resilienz von Gebäuden und Siedlungen im Klimawandel (Stiftung Umwelt und Schadensvorsorge Ed.). München, 12–14.

Schueler, Judith (2008): Materialising Identity. The Co-Construction of the Gotthard Railway and Swiss National Identity. Amsterdam: Aksant.

Schumpeter, Joseph A. (2005): Kapitalismus, Sozialismus und Demokratie. Stuttgart: UTB.

Sikula, Nicole R./Mancillas, James W./Linkov, Igor/McDonagh, John A. (2015): Risk management is not enough: A conceptual model for resilience and adaptation-based vulnerability assessments. In: Environment Systems and Decisions, 35 (2), 219–228.

https://doi.org/10.1007/s10669-015-9552-7

Star, Susan Leigh/Griesemer, James R. (1989): Institutional Ecology, 'Translations' and Boundary Objects. Amateurs and Professionals in Berkeley's Museum of Vertebrate Zoology, 1907–39. In: Social Studies of Science, 19, 387–420.

http://doi.org/10.1177/030631289019003001

Stehrenberger, Cécile (2014): Systeme und Organisationen unter Stress. Zur Geschichte der sozialwissenschaftlichen Katastrophenforschung (1949–1971). In: Zeithistorische Forschungen/ Studies in Contemporary History, 11, 406–424.

Swyngedouw, Eric (2009): The Antinomies of the Postpolitical City: In Search of a Democratic Politics of Environmental Production. In: International Journal of Urban and Regional Research, 33 (3), 601–620.

Bibliography

Theoharidou, M./Kotzanikolaou, P./Gritzalis, D. (2009): Risk-Based Criticality Analysis in Critical Infrastructure Protection III. Berlin, Heidelberg: Springer.

https://doi.org/10.1007/978-3-642-04798-5_3

Turner, Billie Lee (2010): Vulnerability and Resilience. Coalescing or Paralleling Approaches for Sustainability Science? In: Global Environmental Change, 20 (4), 570–576.

https://doi.org/10.1016/j.gloenvcha.2010.07.003

Turner, Billie Lee/Kasperson, Roger E./Matson, Pamela A./McCarthy, James J./Corell, Robert W./Christensen, Lindsey/Eckley, Noelle/Kasperson, Jeanne X./Luers, Amy/Martello, Marybeth L./Polsky, Colin/Pulsipher, Alexander/Schiller, Andrew (2003): A Framework for Vulnerability Analysis in Sustainability Science. In: PNAS, 100 (14), 8074–8079.

https://doi.org/10.1073/pnas.1231335100

UN-Habitat (2017): City Resilience Profiling Programme. Nairobi: UN-Habitat.

https://unhabitat.org/urban-initiatives/initiatives-programmes/city-resilience-profiling-programme

UNISDR (2004): Living with Risk. A global review of disaster reduction initiatives. Geneva: United Nations.

UNISDR (2009): Terminology on Disaster Risk Reduction. United Nations International Strategy for Disaster Reduction. Geneva: United Nations.

United States Government (2003): The National Strategy for the Physical Protection of Critical Infrastructures and Key Assets. Washington.

Vleuten, Erik van der (2006): Understanding Network Societies. Two Decades of Large Technical System Studies. In: Vleuten, Erik van der/ Kaijser, Arne (Eds.): Networking Europe. Transnational Infrastructure and the Shaping of Europe 1850–2000. Sagamore Beach: Science History Publications, 279–314.

Vleuten, Erik van der (2017): Challenging Prometheus. A history of technology for an age of grand challenges. Eindhoven: University of Eindhoven.

Vugrin, Eric D./Warren, Drake E./Ehlen, Mark A./Camphouse, R. Chris (2010): A Framework for Assessing the Resilience of Infrastructure and Economic Systems. In: Gopalakrishnan, Kasthurirangan/Peeta, Srinivas (Eds.): Sustainable and Resilient Critical Infrastructure Systems. Heidelberg: Springer, 77–116.

Watts, Michael J./Bohle, Hans G. (1993): The Space of Vulnerability. The Causal Structure of Hunger and Famine. In: Progress in Human Geography, 17 (1), 43–67.

http://doi.org/10.1177/030913259301700103

White, Iain/O'Hare, Paul (2014): From Rhetoric to Reality: Which Resilience, Why Resilience, and Whose Resilience in Spatial Planning? In: Environment and Planning C: Government and Policy, 32 (5), 934–950.

http://doi.org/10.1068/c12117?id=c12117

White House (1998): The Clinton Administration's Policy on Critical Infrastructure Protection: Presidential Decision Directive 63. nsc-63 pdd §. Washington.

Wisner, Ben/O'Keefe, Phil/Westgate, Ken (1977): Global Systems and Local Disasters. The Untapped Power of People's Science. In: Disasters, 1 (1), 47–57.

https://doi.org/10.1111/j.1467-7717.1977.tb00008.x

Wisner, Ben/Blaikie, Piers/Cannon, Terry/Davis, Ian (2003): At Risk. Natural Hazards, People's Vulnerability and Disasters. London/New York: Routledge.

Zebrowski, Chris (2013): The nature of resilience. In: Resilience, 1 (3), 159–173.
https://doi.org/10.1080/21693293.2013.804672

GPSR Compliance
The European Union's (EU) General Product Safety Regulation (GPSR) is a set
of rules that requires consumer products to be safe and our obligations to
ensure this.

If you have any concerns about our products, you can contact us on

ProductSafety@springernature.com

In case Publisher is established outside the EU, the EU authorized
representative is:

Springer Nature Customer Service Center GmbH
Europaplatz 3
69115 Heidelberg, Germany